Hello, happy to see you here.

I'm BB, aka Bead Baby.

Abacus is fun, if you know All About it.

"How to Abacus" has books to help you.

Book 1: All About Abacus

Book 2: All About Formulas

Book 3: All About Multiplication

Book 4: All About Division

I will definitely be with you through all the lessons.

Let's go!

AF887771

To those interested in Abacus

~Sheena

A great book for beginners

~Yuenjo

For information address PinGrow Media, 1049 El Monte Avenue Ste C # 586, Mountain View California 94040 or contact@pingrow.com

ISBN: 978-1-949622-04-1

Visit www.pingrow.com

How to Abacus 3

All About Multiplication

Multiplication and Abacus

Multiplication and Abacus

Multiplication problems contain a serial multiplying process and each result of the multiplication operation, also known as the product, needs to be added together.

When solving multiplication questions with an abacus, it is necessary to establish the fundamental knowledge by reviewing the other books in the series: How to Abacus 1 & How to Abacus 2.

1. Formulas

Numbers need to be added on during the multiplying process and formulas are the keys to solve any additions on the abacus. Therefore, it is crucial to be very familiar with the formulas of Little Friends, Big Friends and Mixed Friends that are introduced in “How to Abacus 2”.

Big Friends Formulas

+ 1 = - 9 + 10	- 1 = - 10 + 9
+ 2 = - 8 + 10	- 2 = - 10 + 8
+ 3 = - 7 + 10	- 3 = - 10 + 7
+ 4 = - 6 + 10	- 4 = - 10 + 6
+ 5 = - 5 + 10	- 5 = - 10 + 5
+ 6 = - 4 + 10	- 6 = - 10 + 4
+ 7 = - 3 + 10	- 7 = - 10 + 3
+ 8 = - 2 + 10	- 8 = - 10 + 2
+ 9 = - 1 + 10	- 9 = - 10 + 1

Little Friends Formulas

+ 1 = + 5 – 4	- 1 = - 5 + 4
+ 2 = + 5 – 3	- 2 = - 5 + 3
+ 3 = + 5 – 2	- 3 = - 5 + 2
+ 4 = + 5 – 1	- 4 = - 5 + 1

Mixed Friends Formulas

+ 6 = - 5 + 1 + 10	- 6 = - 10 + 5 -1
+ 7 = - 5 + 2 + 10	- 7 = - 10 + 5 - 2
+ 8 = - 5 + 3 + 10	- 8 = - 10 + 5 - 3
+ 9 = - 5 + 4 + 10	- 9 = - 10 + 5 - 4

2. Times Table

Multiplication is a simplified form of additions. No matter how many digits a multiplication problem would involve, the algorithm process would still go step by step, meaning each digit would multiply another digit accordingly.

Using the Times Table is the fastest method to find the product for any multiplication operation. It is essential to memorize the Times Table when solving multiplication problems on the abacus.

Abacus Fun Fact:

The Chinese abacus or the "suanpan" was first used thousands of years ago in the 2nd century BC.

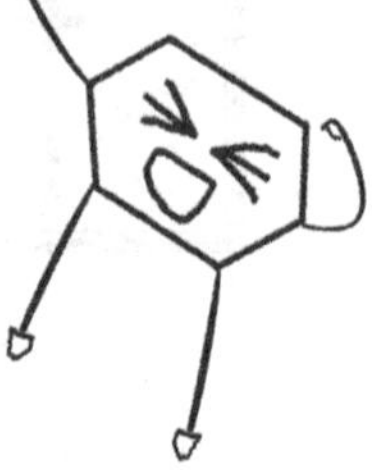

Times Table

2 x 1 = 02	3 x 1 = 03	4 x 1 = 04	5 x 1 = 05
2 x 2 = 04	3 x 2 = 06	4 x 2 = 08	5 x 2 = 10
2 x 3 = 06	3 x 3 = 09	4 x 3 = 12	5 x 3 = 15
2 x 4 = 08	3 x 4 = 12	4 x 4 = 16	5 x 4 = 20
2 x 5 = 10	3 x 5 = 15	4 x 5 = 20	5 x 5 = 25
2 x 6 = 12	3 x 6 = 18	4 x 6 = 24	5 x 6 = 30
2 x 7 = 14	3 x 7 = 21	4 x 7 = 28	5 x 7 = 35
2 x 8 = 16	3 x 8 = 24	4 x 8 = 32	5 x 8 = 40
2 x 9 = 18	3 x 9 = 27	4 x 9 = 36	5 x 9 = 45

6 x 1 = 06	7 x 1 = 07	8 x 1 = 08	9 x 1 = 09
6 x 2 = 12	7 x 2 = 14	8 x 2 = 16	9 x 2 = 18
6 x 3 = 18	7 x 3 = 21	8 x 3 = 24	9 x 3 = 27
6 x 4 = 24	7 x 4 = 28	8 x 4 = 32	9 x 4 = 36
6 x 5 = 30	7 x 5 = 35	8 x 5 = 40	9 x 5 = 45
6 x 6 = 36	7 x 6 = 42	8 x 6 = 48	9 x 6 = 54
6 x 7 = 42	7 x 7 = 49	8 x 7 = 56	9 x 7 = 63
6 x 8 = 48	7 x 8 = 56	8 x 8 = 64	9 x 8 = 72
6 x 9 = 54	7 x 9 = 63	8 x 9 = 72	9 x 9 = 81

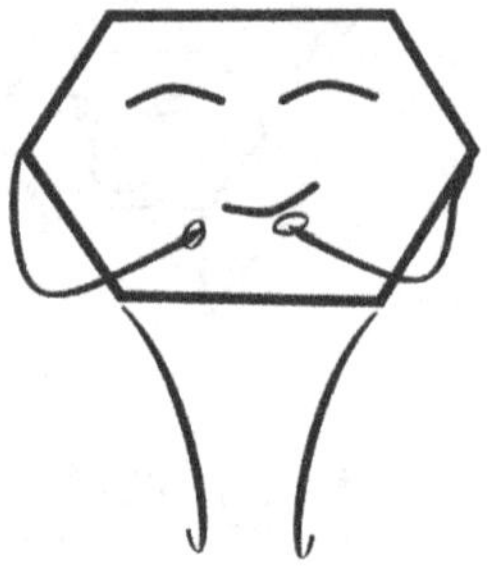

3. Leave a space for zero

While using an abacus for multiplication arithmetic, every number is important, including zero. A space for zero should be maintained on the abacus. If the product is one digit during the operation, then the product should have a zero before that one digit such as 02, 03, 04 and so on.

The reason to leave a space for zero during the operation is that abacus arithmetic does not start from the ones place. Abacus algorithm always starts from left to right and puts the answer from left to right on the abacus accordingly.

Multiplication problems usually contain multiple algorithms; therefore, it is important to leave a space for zero on the abacus when the product is one digit during the operation.

4. The starting place

For addition and subtraction questions, there is no need to find the starting place to solve problems. You just follow the numbers on the questions to do the algorithm. However, for multiplication questions, it is vital to know where to put the answers on the abacus.

Multiplication questions involve serial multiplying and adding algorithms. You need to know the starting place for where to put the series of products on the abacus and add these products together for the final answer. There is a starting place to put the product for each digit multiplying another digit.

How to find the starting place

The key point for finding the starting place is to add the digits of the two factors you are multiplying.

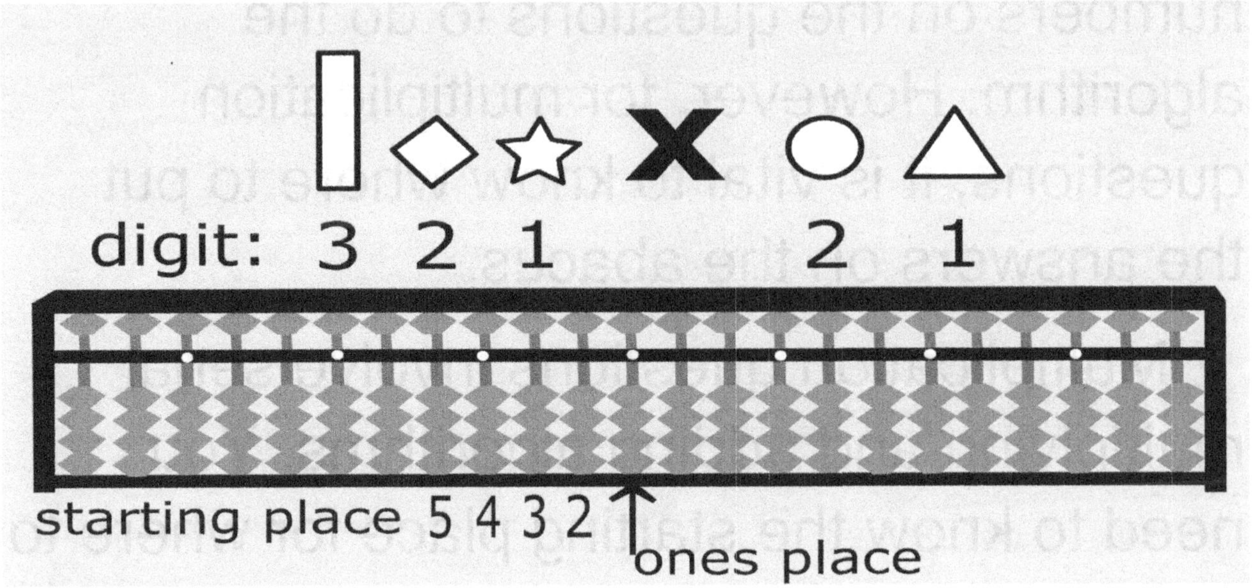

The diagram above shows that ▯◇☆, a three-digit number, multiplies ○△, a two-digit number. In the number of ▯◇☆, ▯ is the 3rd digit, ◇ is the 2nd digit and ☆ is the 1st digit. In the number of ○△, ○ is the 2nd digit and △ is the 1st digit.

Starting place = sum of the digits

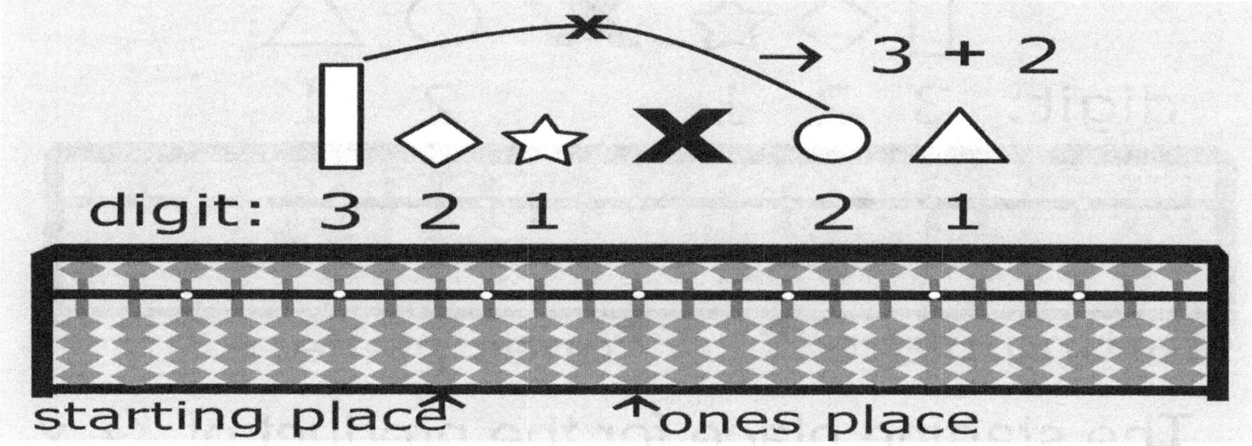

The starting place for the product of ▯ x ○ is the 5th place on the abacus because 3 digits (▯)+ 2 digits (○) = 5 digits.

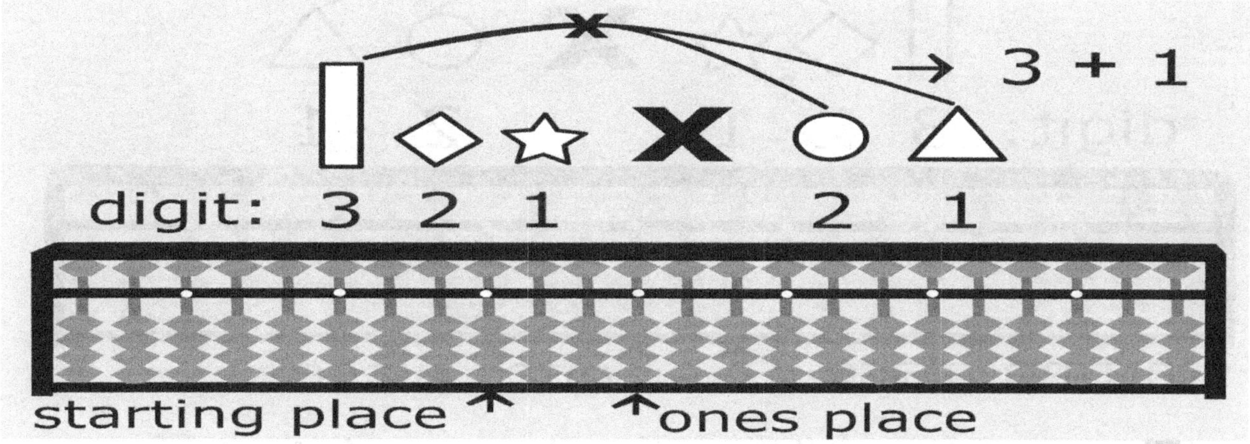

The starting place for the product of ▯ x △ is the 4th place on the abacus because 3 digits (▯)+ 1 digits (△) = 4 digits.

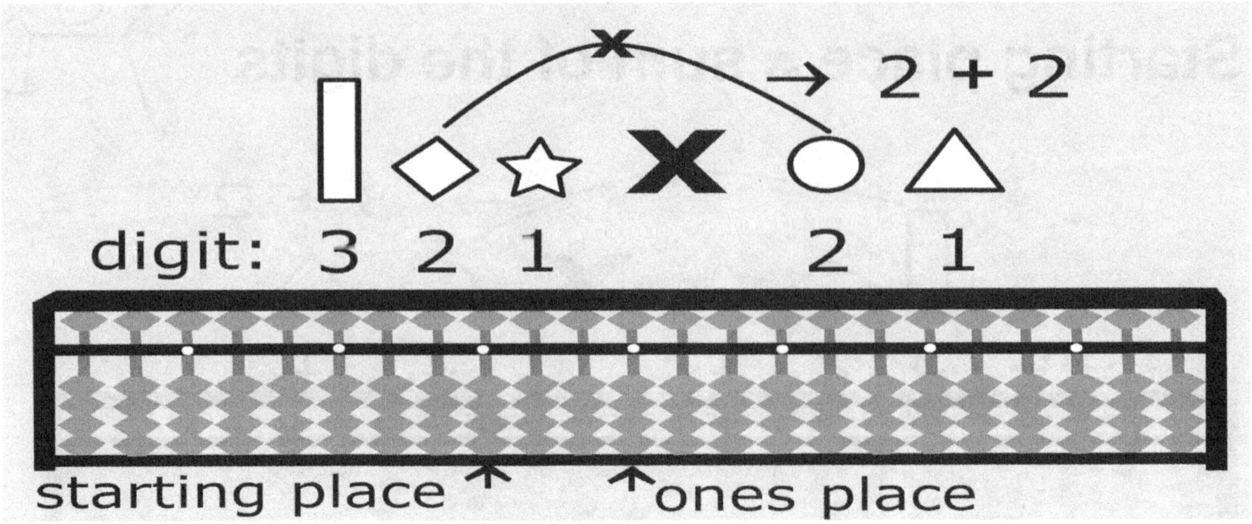

The starting place for the product of ◇ x ○ is the 4th place on the abacus because 2 digits (◇) + 2 digits (○) = 4 digits.

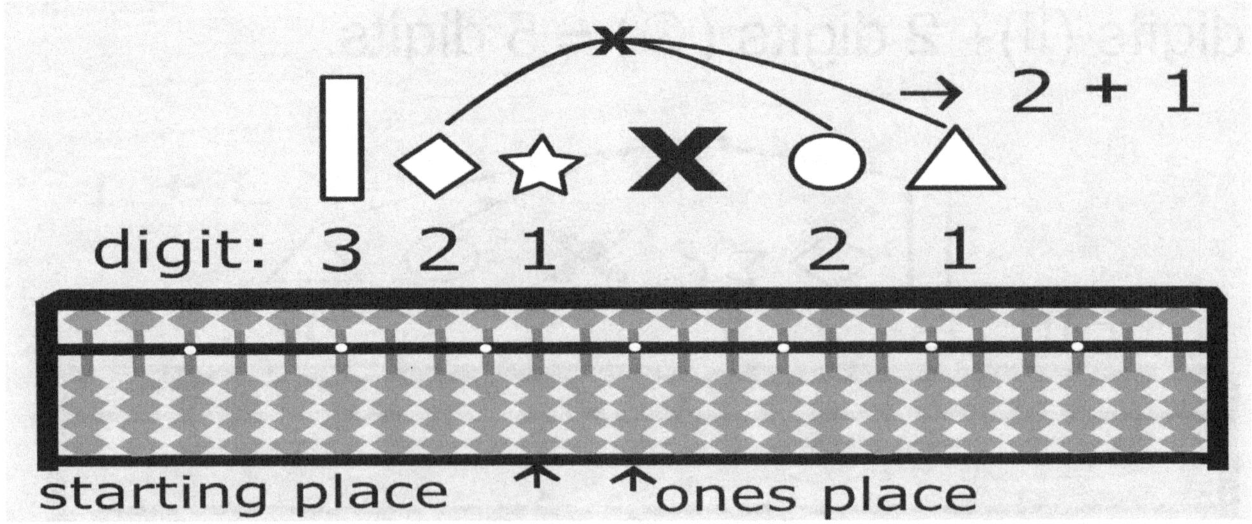

The starting place for the produt of ◇ x △ is the 3rd place on the abacus because 2 digit (◇) + 1 digit (△) = 3 digits.

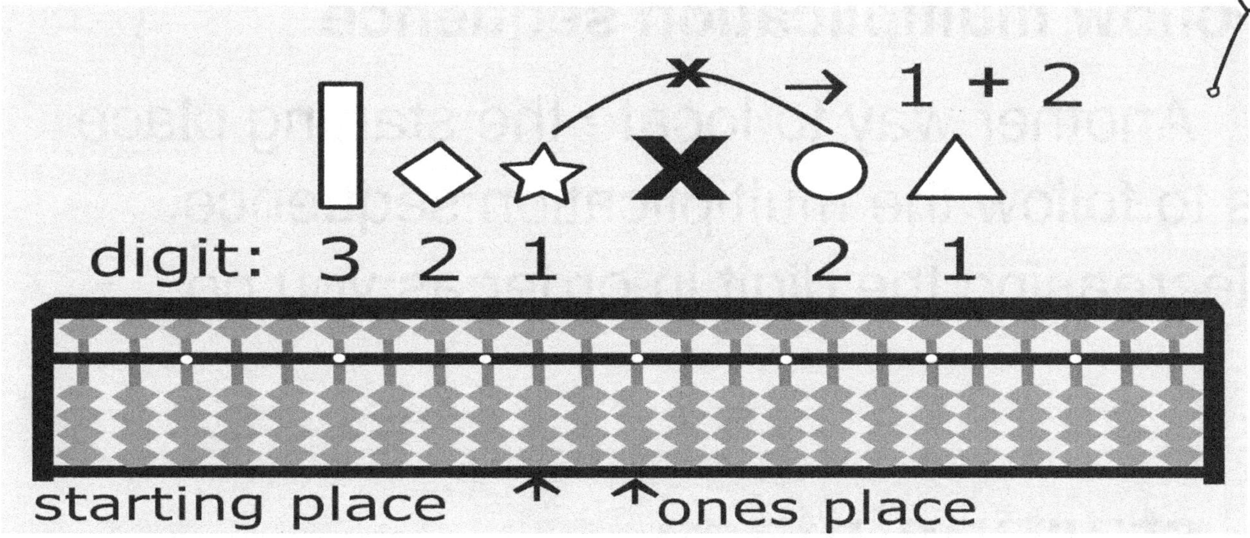

The starting place for the produt of ☆ x ○ is the 3rd place on the abacus because 1 digit (☆) + 2 digit (○) = 3 digits.

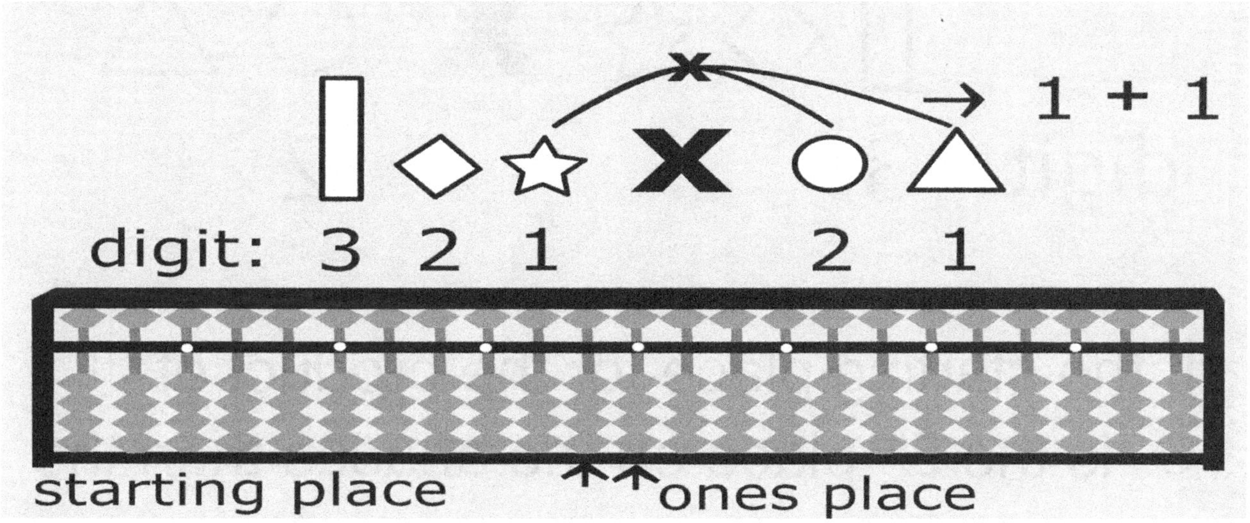

The starting place for the produt of ☆ x △ is the 2nd place on the abacus because 1 digit (☆) + 1 digit (△) = 2 digits.

Follow multiplication sequence

Another way to locate the starting place is to follow the multiplication sequence, decreasing the digit in order as you go.

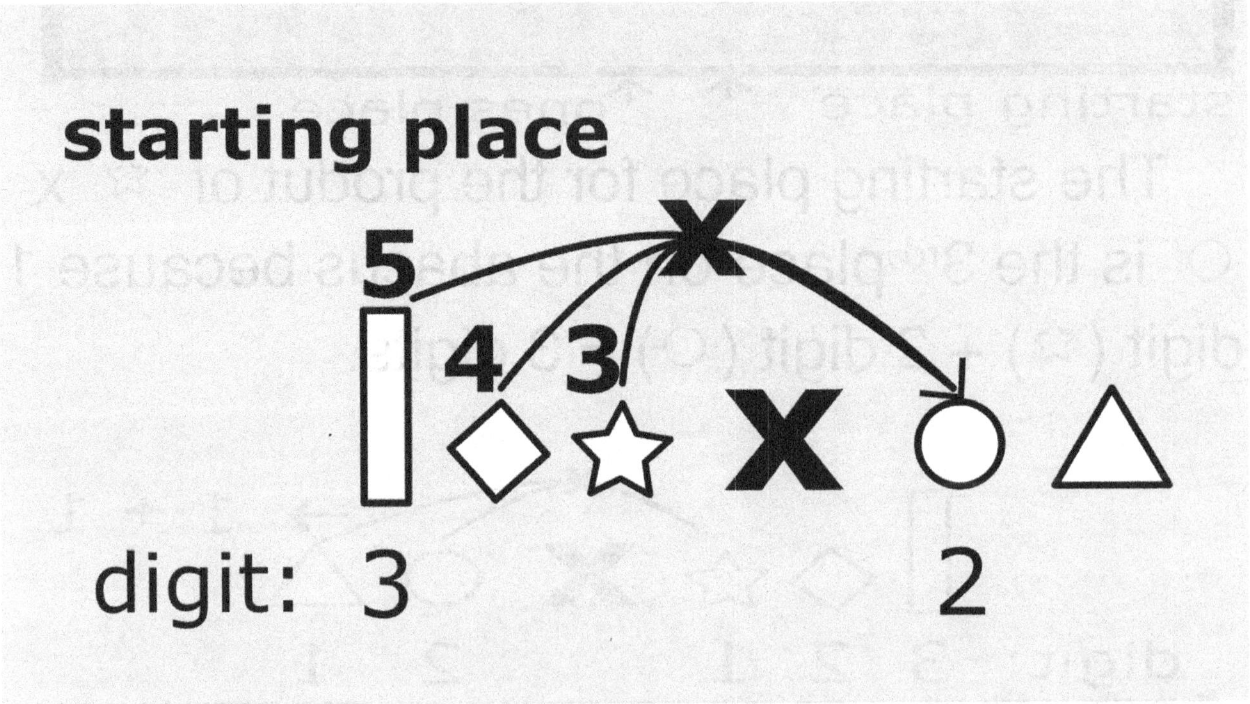

If the starting place for the product of ▯ x ○ is the 5th place on the abacus then the starting place for the product of ◇ x ○ is the 4th place, and the product of ☆ x ○ is the 3rd place on the abacus.

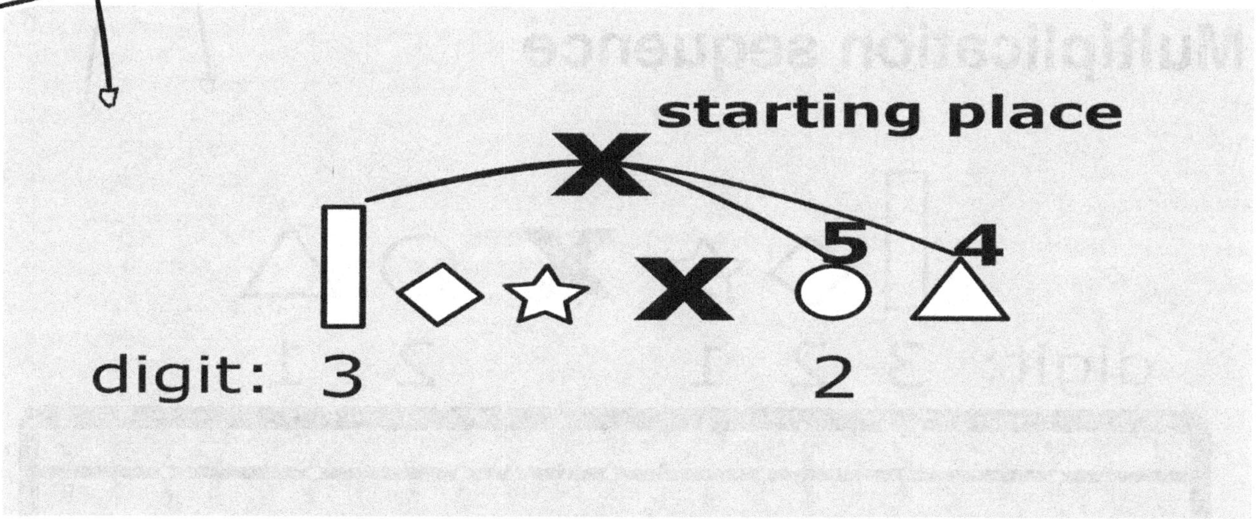
starting place
5
4
digit: 3
2

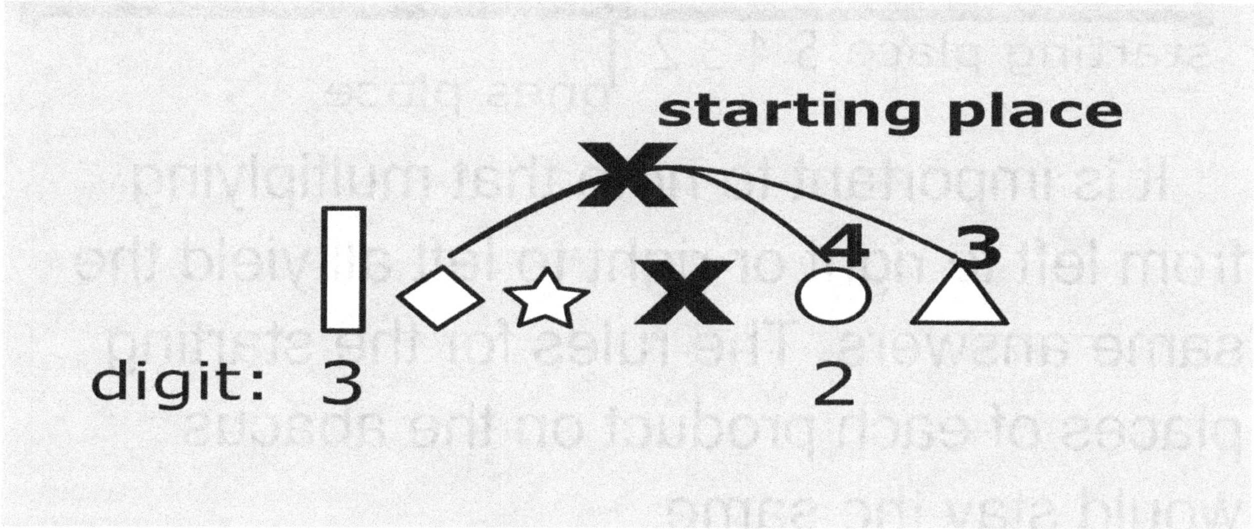
starting place
4
3
digit: 3
2

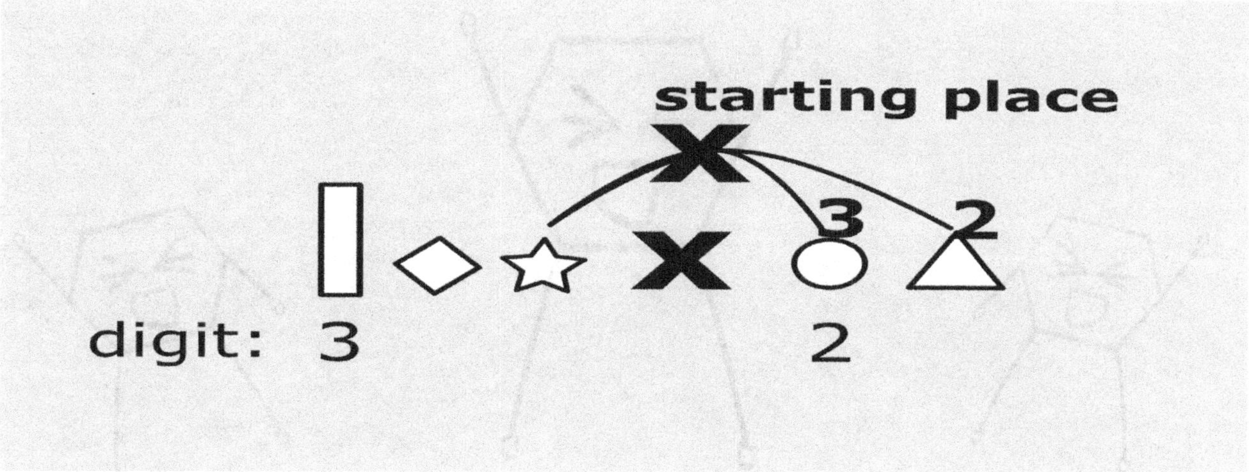
starting place
3
2
digit: 3
2

Multiplication sequence

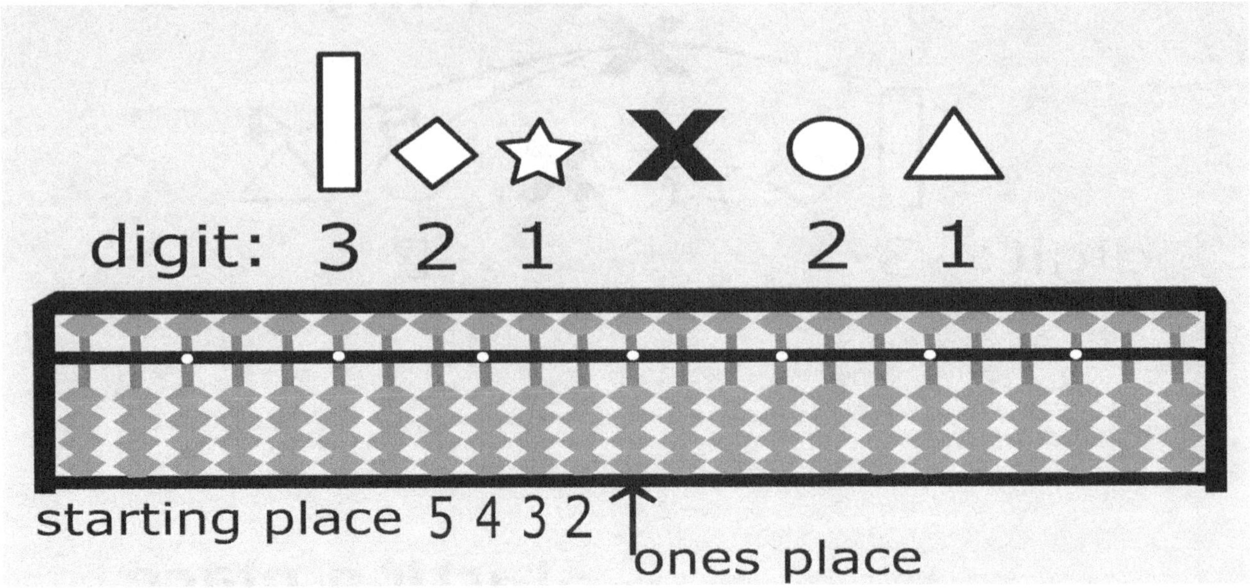

It is important to note that multiplying from left to right or right to left all yield the same answers. The rules for the starting places of each product on the abacus would stay the same.

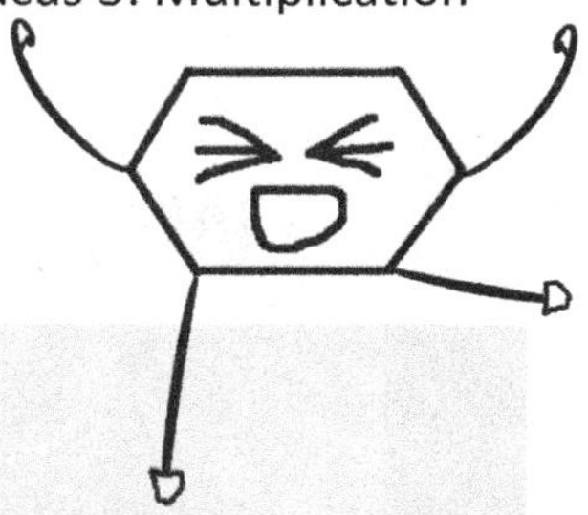

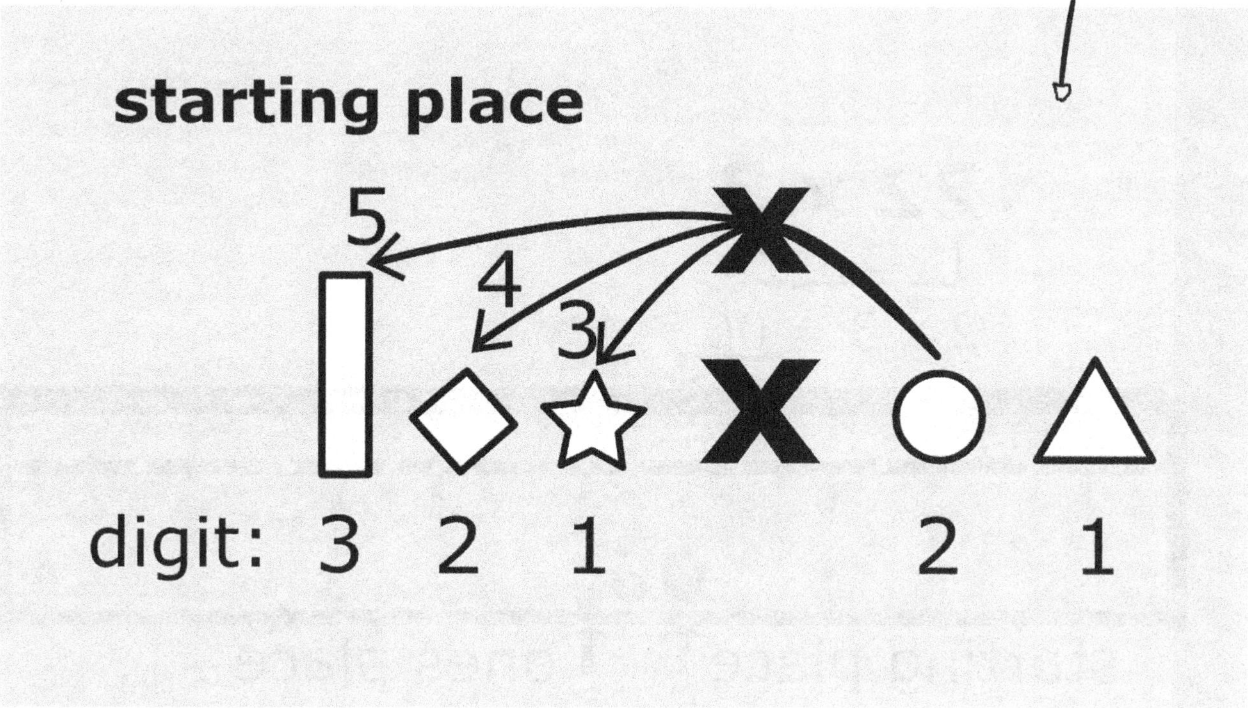
starting place
5
4
3
digit: 3 2 1 2 1

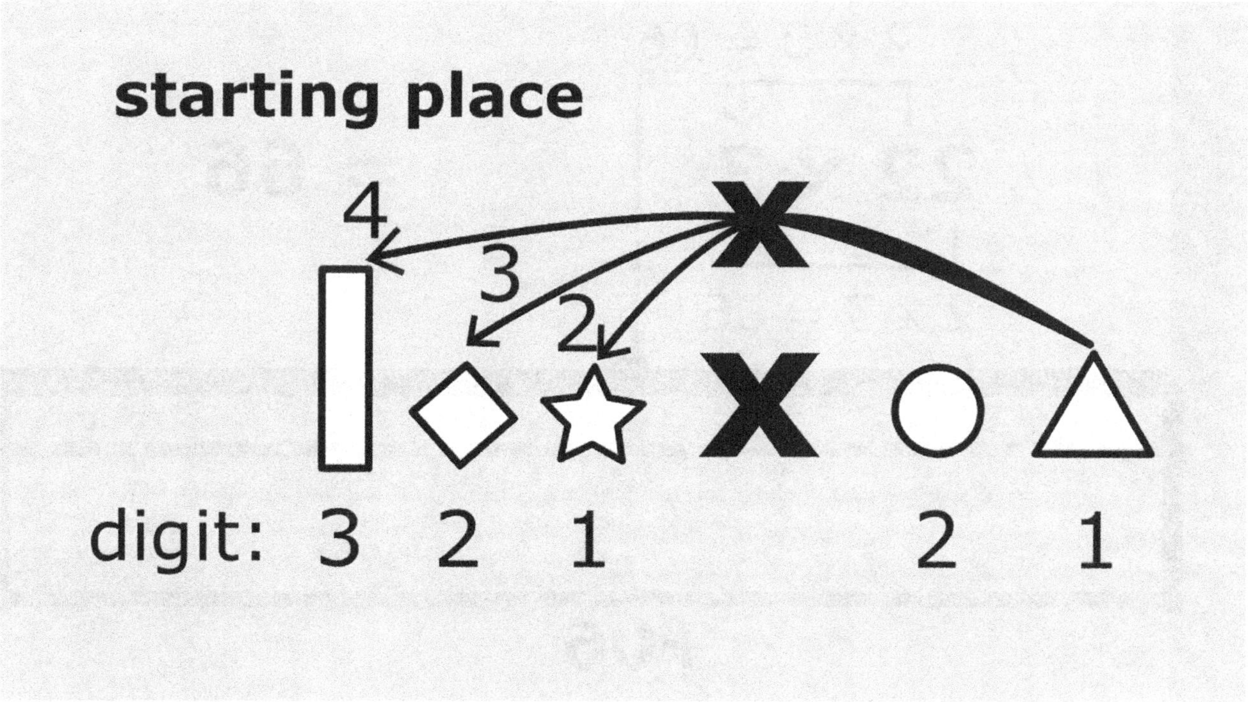
starting place
4
3
2
digit: 3 2 1 2 1

5. two digits times one digit

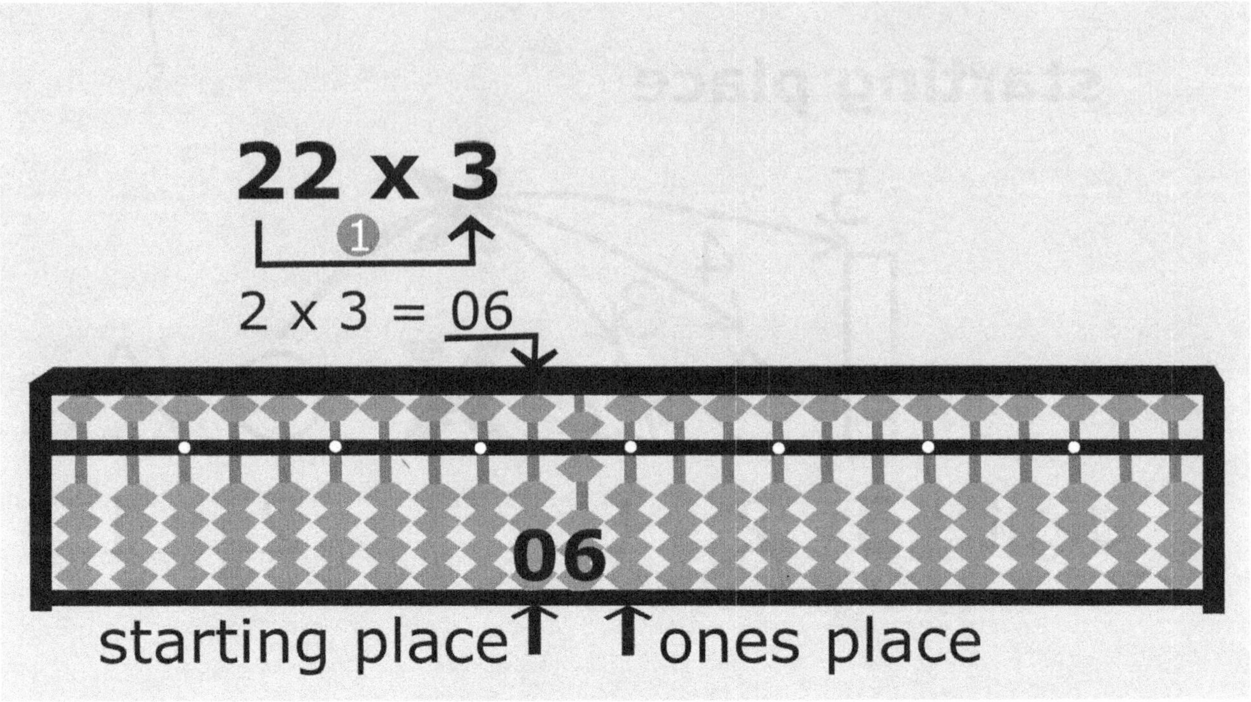

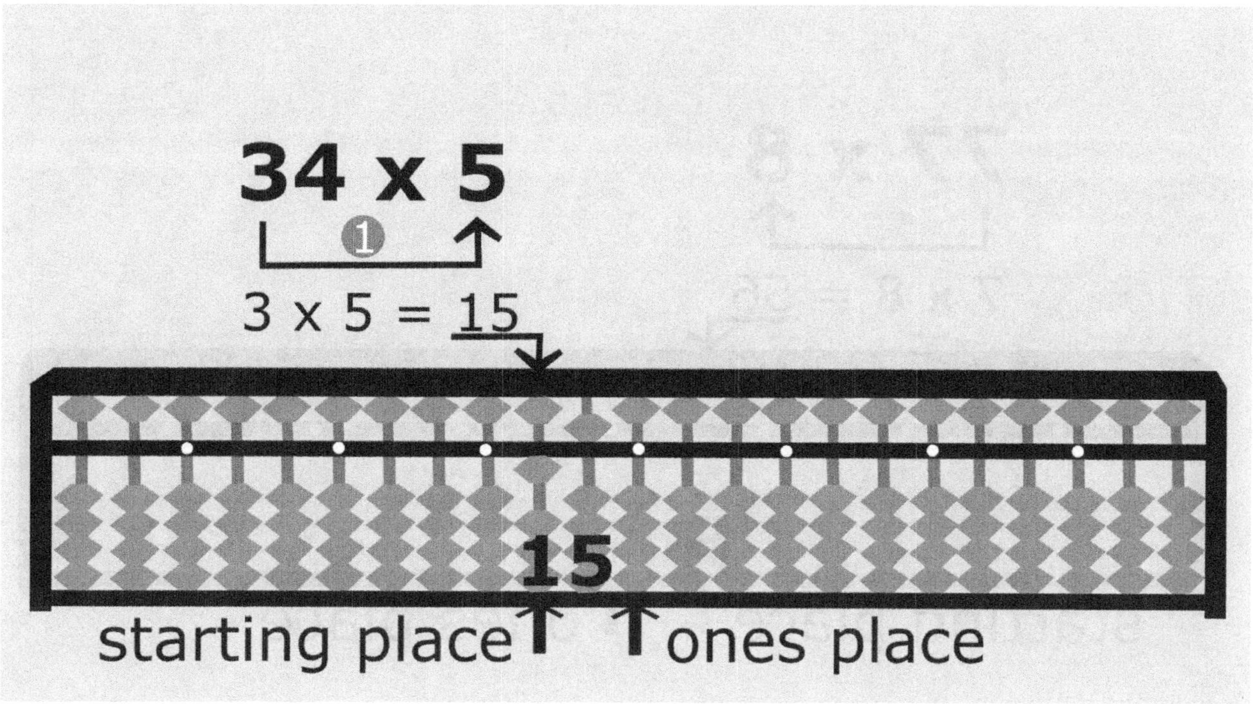
34 x 5
1
3 x 5 = 15
15
starting place
ones place

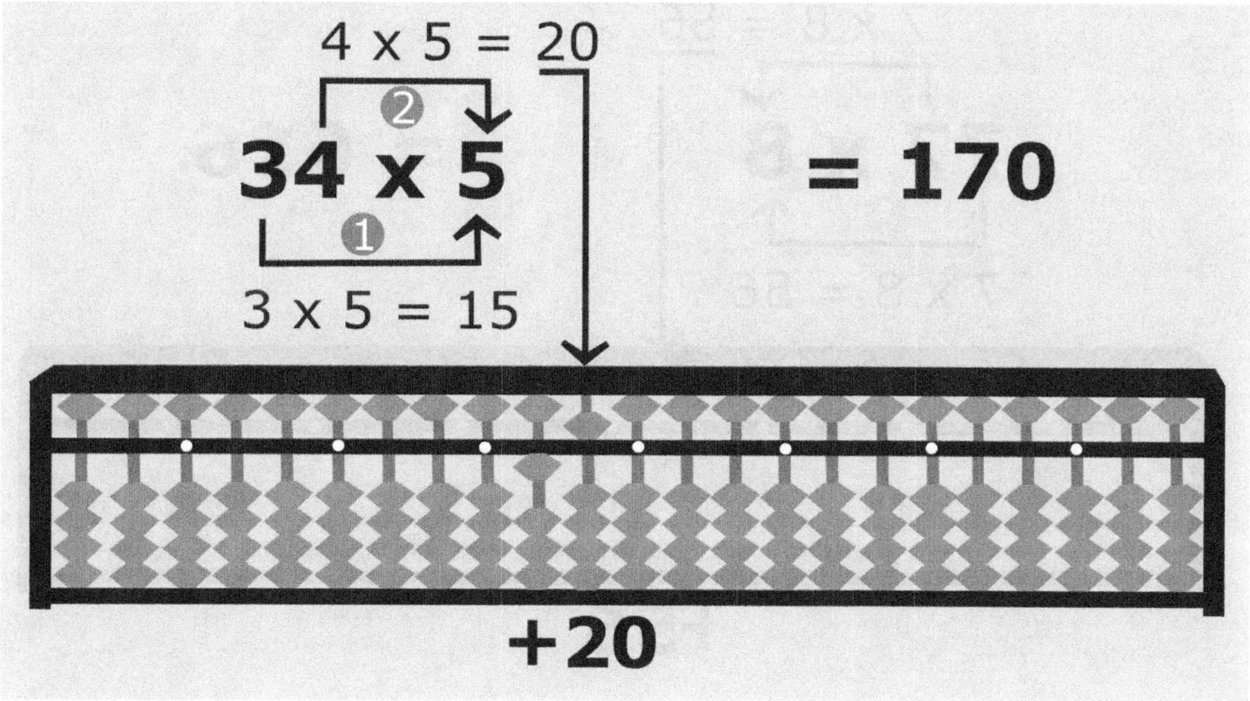
4 x 5 = 20
2
34 x 5
1
3 x 5 = 15
= 170
+20

Go! Go!Go!

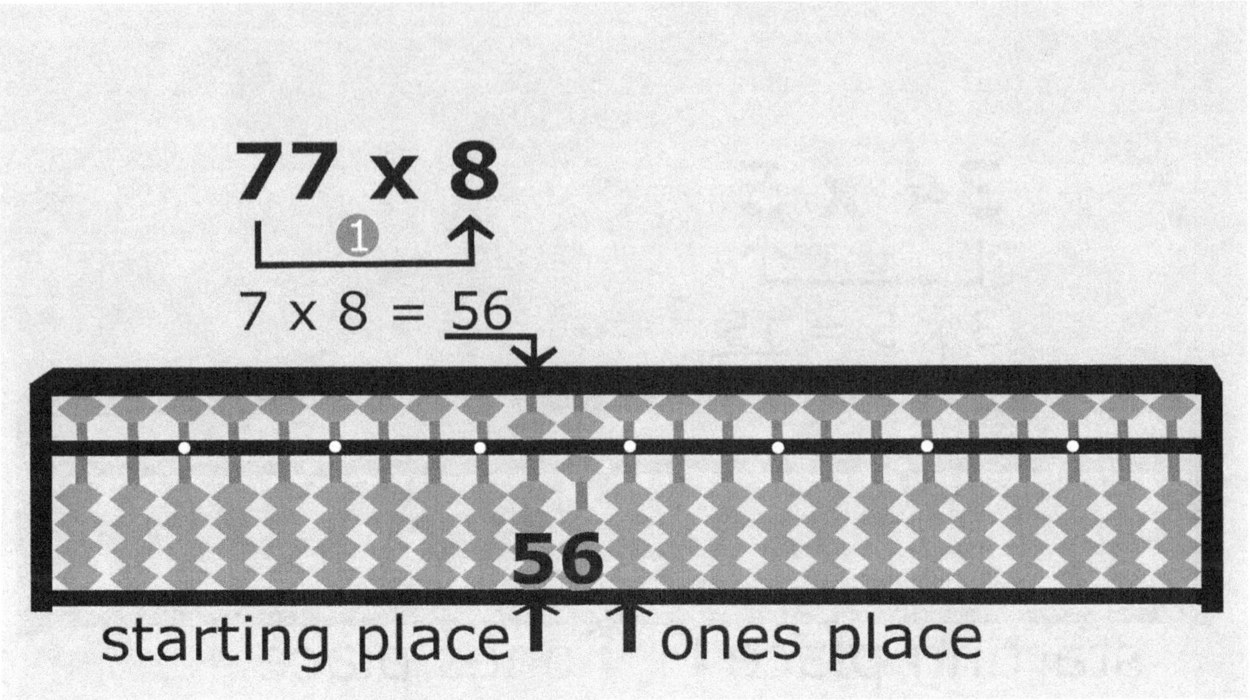
77 x 8
1
7 x 8 = 56
56
starting place
ones place

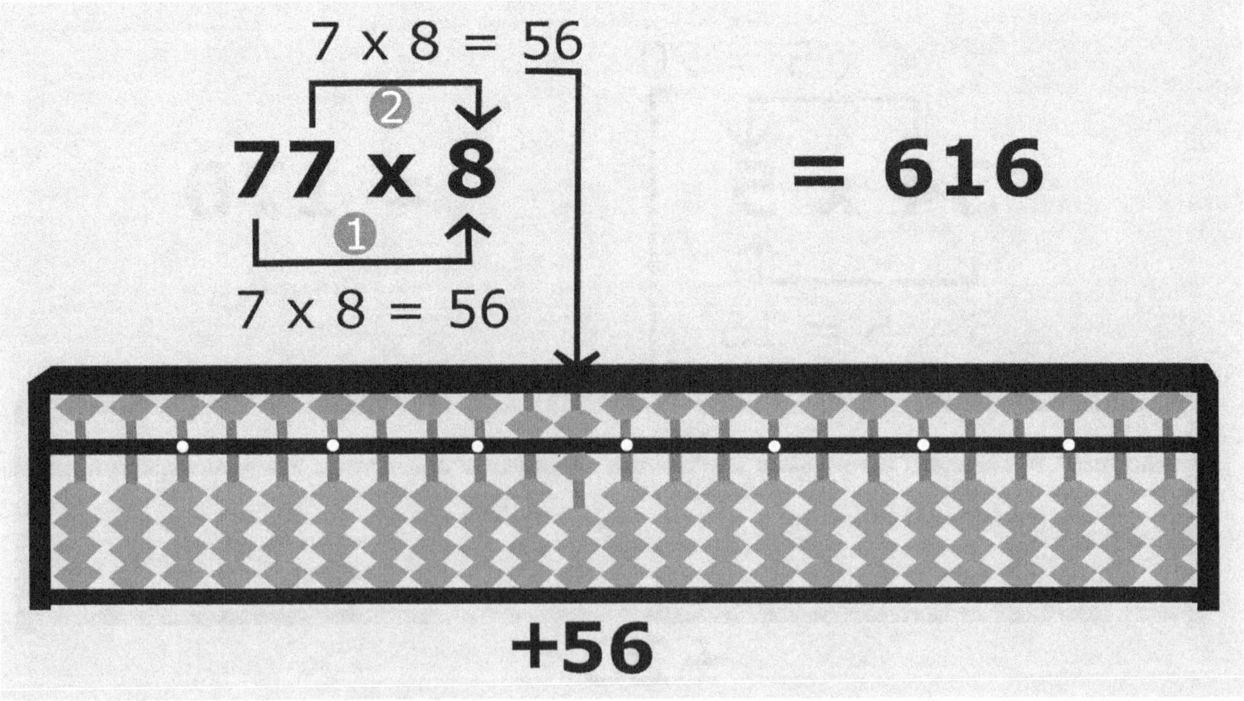
7 x 8 = 56
2
77 x 8
1
7 x 8 = 56
= 616
+56

Practice for 2 digits times 1 digit

1) 46 X 6 =	15) 52 X 4 =
2) 98 X 6 =	16) 49 X 9 =
3) 65 X 8 =	17) 93 X 5 =
4) 81 X 7 =	18) 25 X 9 =
5) 42 X 6 =	19) 69 X 8 =
6) 22 X 9 =	20) 94 X 2 =
7) 64 X 8 =	21) 58 X 5 =
8) 74 X 3 =	22) 38 X 5 =
9) 58 X 4 =	23) 45 X 8 =
10) 12 X 4 =	24) 47 X 8 =
11) 87 X 4 =	25) 35 X 6 =
12) 38 X 9 =	26) 99 X 8 =
13) 76 X 6 =	27) 94 X 8 =
14) 79 X 4 =	28) 69 X 3 =

6. three digits times one digit

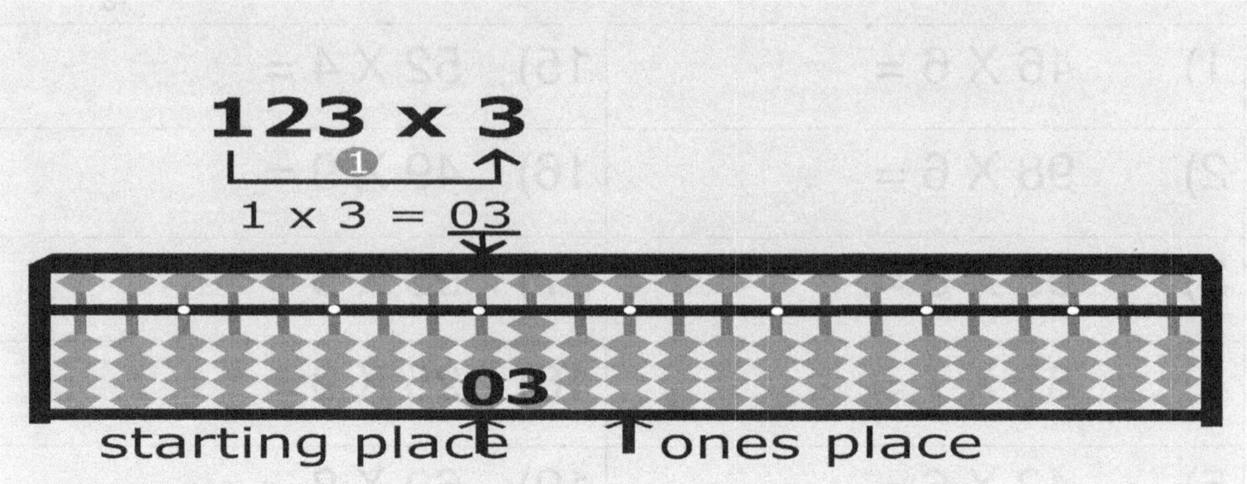

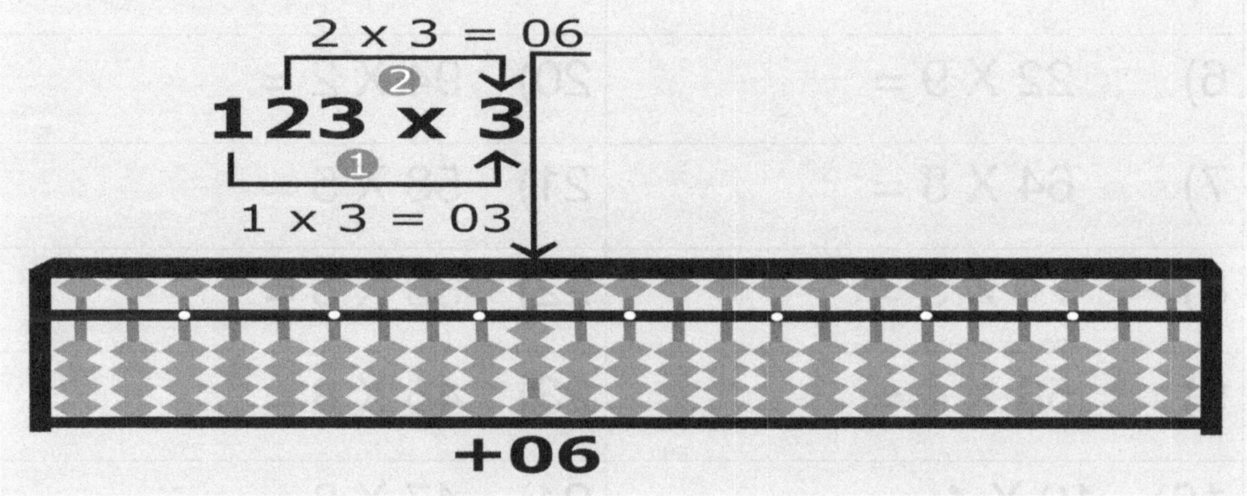

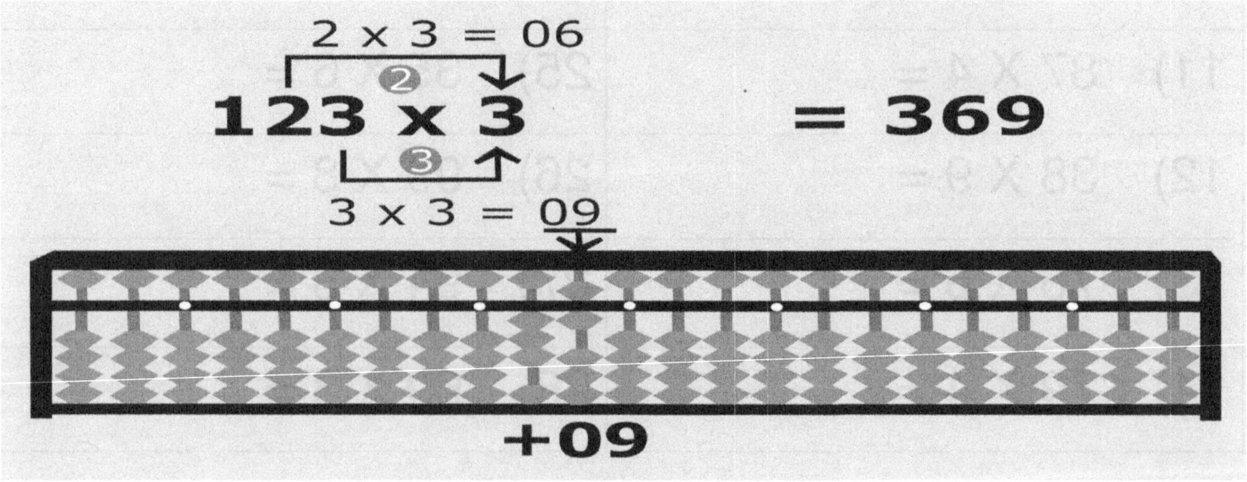

Go! Go!Go!

Go! Go!Go!

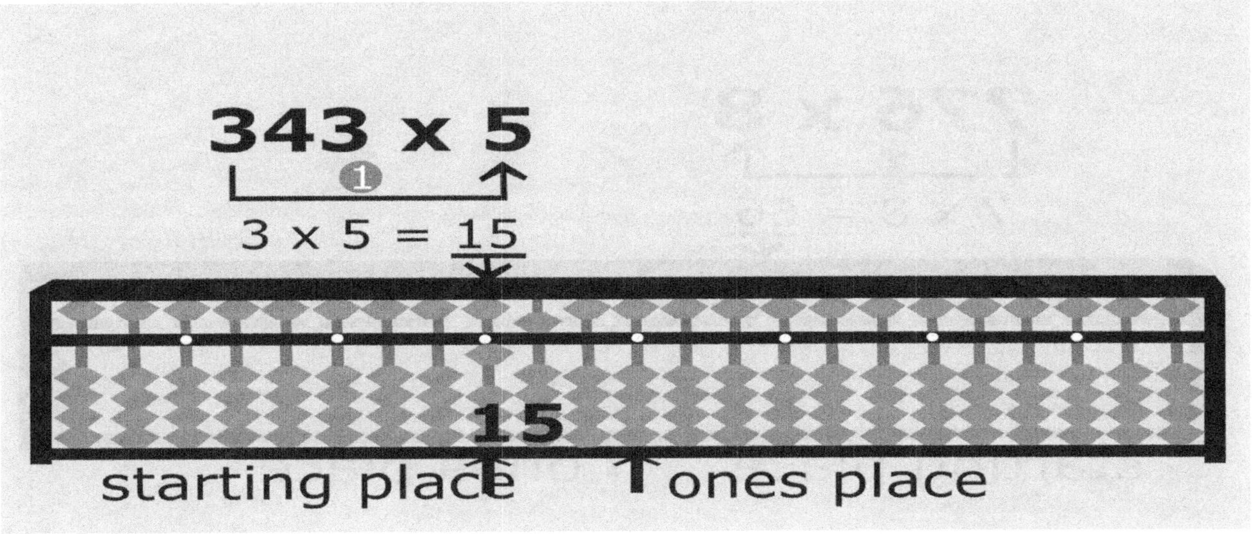
343 x 5
1
3 x 5 = 15
15
starting place
ones place

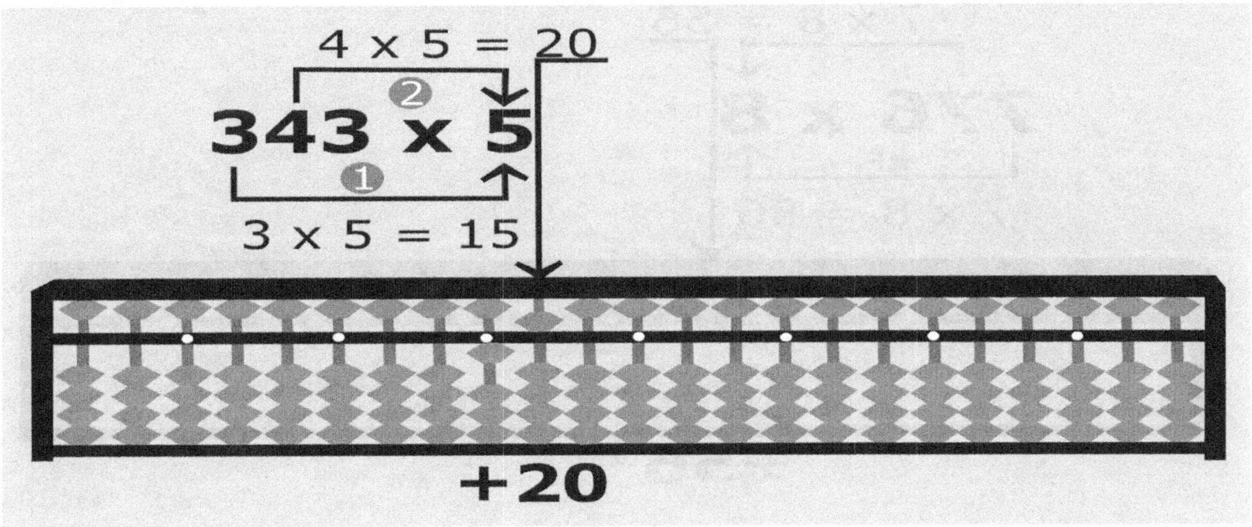
4 x 5 = 20
2
343 x 5
1
3 x 5 = 15
+20

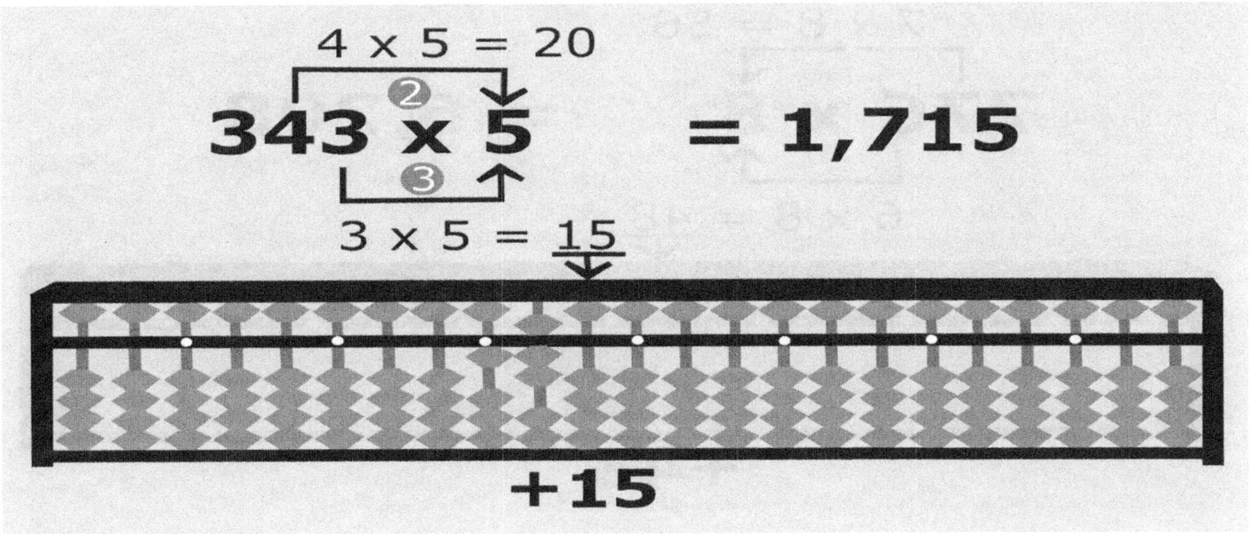
4 x 5 = 20
2
343 x 5
= 1,715
3
3 x 5 = 15
+15

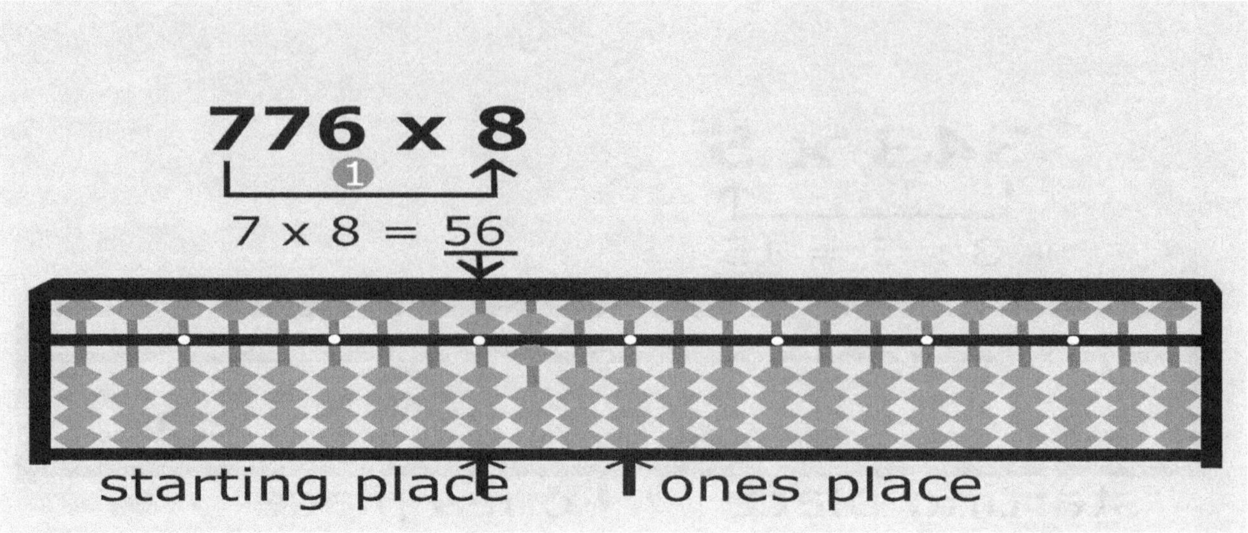
776 x 8
1
7 x 8 = 56
starting place
ones place

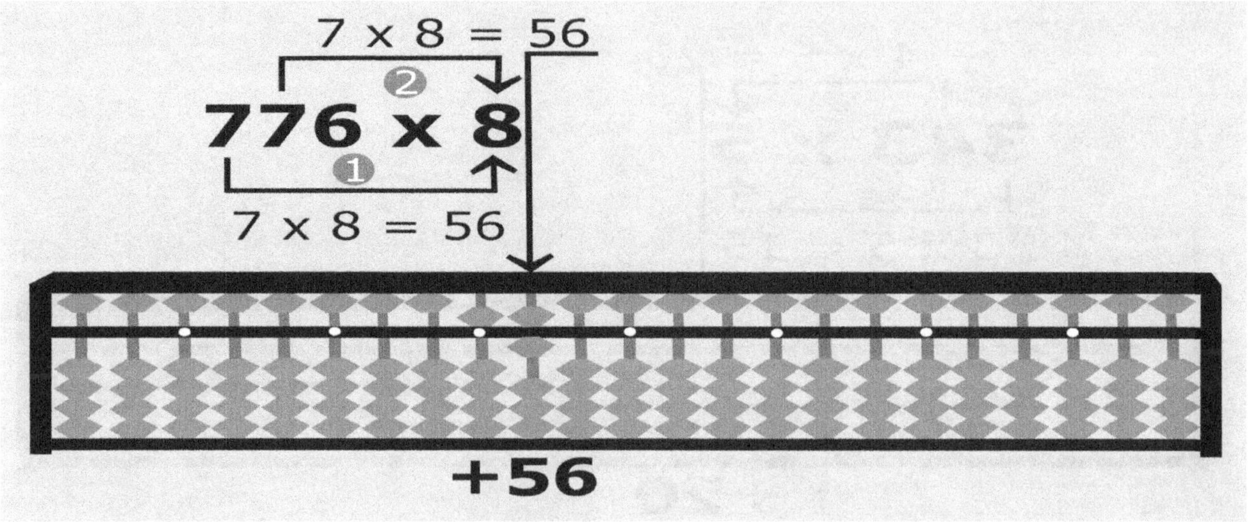
7 x 8 = 56
2
776 x 8
1
7 x 8 = 56
+56

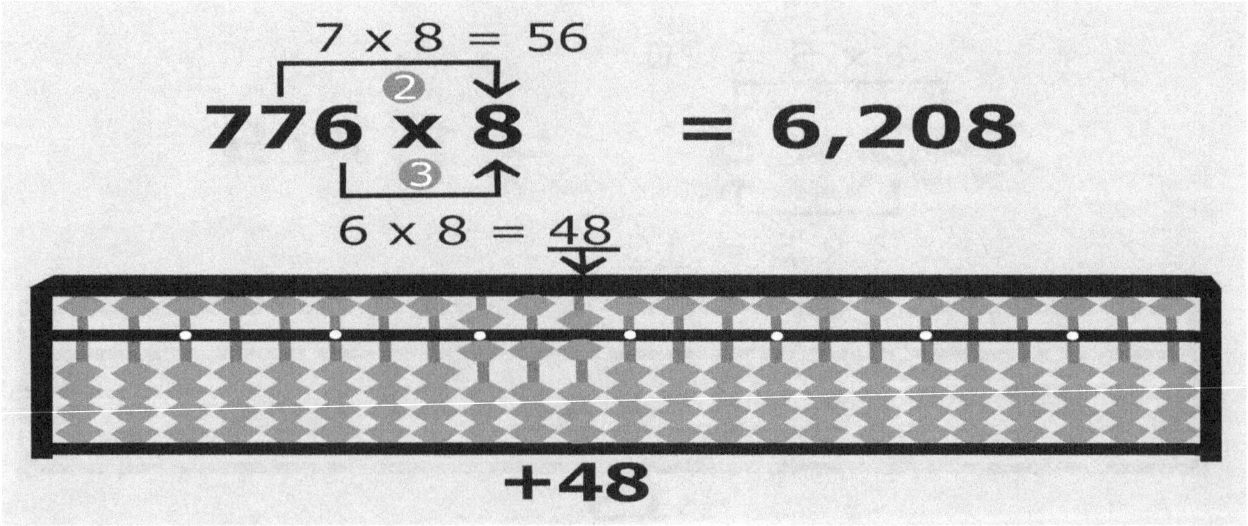
7 x 8 = 56
2
776 x 8
= 6,208
3
6 x 8 = 48
+48

Practice for 3 digits times 1 digit

1) 959 X 6 =	15) 247 X 7 =
2) 483 X 2 =	16) 965 X 6 =
3) 862 X 6 =	17) 361 X 4 =
4) 321 X 7 =	18) 533 X 9 =
5) 524 X 2 =	19) 682 X 5 =
6) 214 X 4 =	20) 933 X 6 =
7) 349 X 9 =	21) 190 X 3 =
8) 733 X 8 =	22) 831 X 4 =
9) 442 X 6 =	23) 609 X 4 =
10) 153 X 9 =	24) 185 X 5 =
11) 255 X 8 =	25) 559 X 3 =
12) 138 X 6 =	26) 892 X 6 =
13) 565 X 3 =	27) 685 X 6 =
14) 181 X 5 =	28) 287 X 8 =

7. two digits times two digits

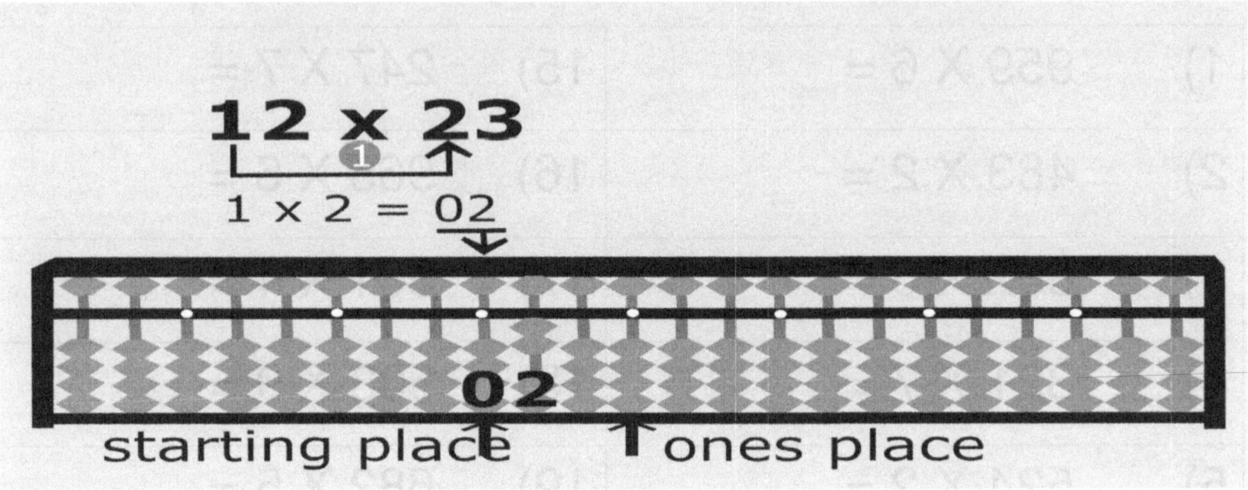

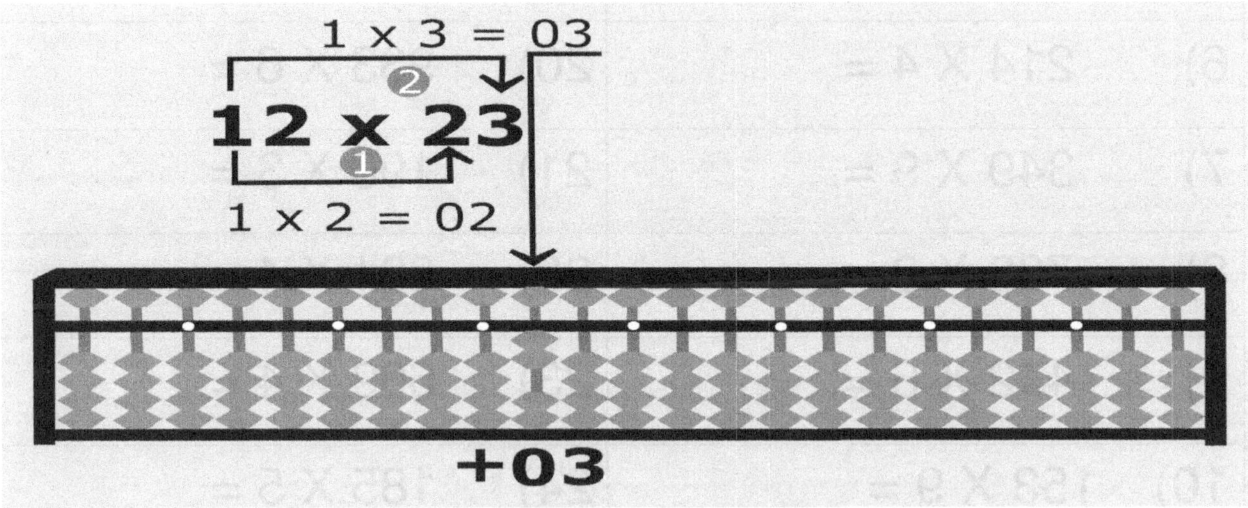

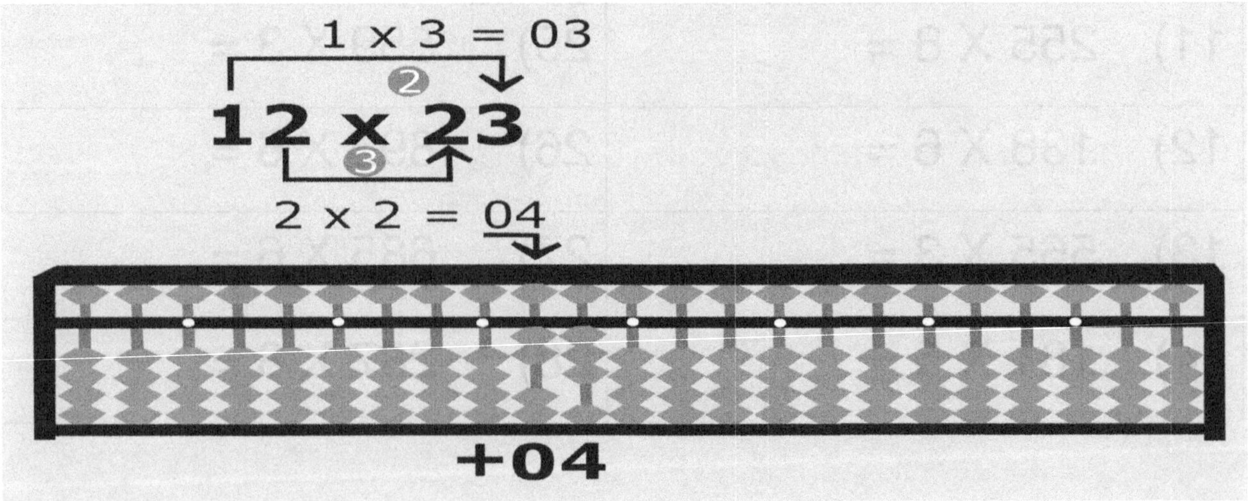

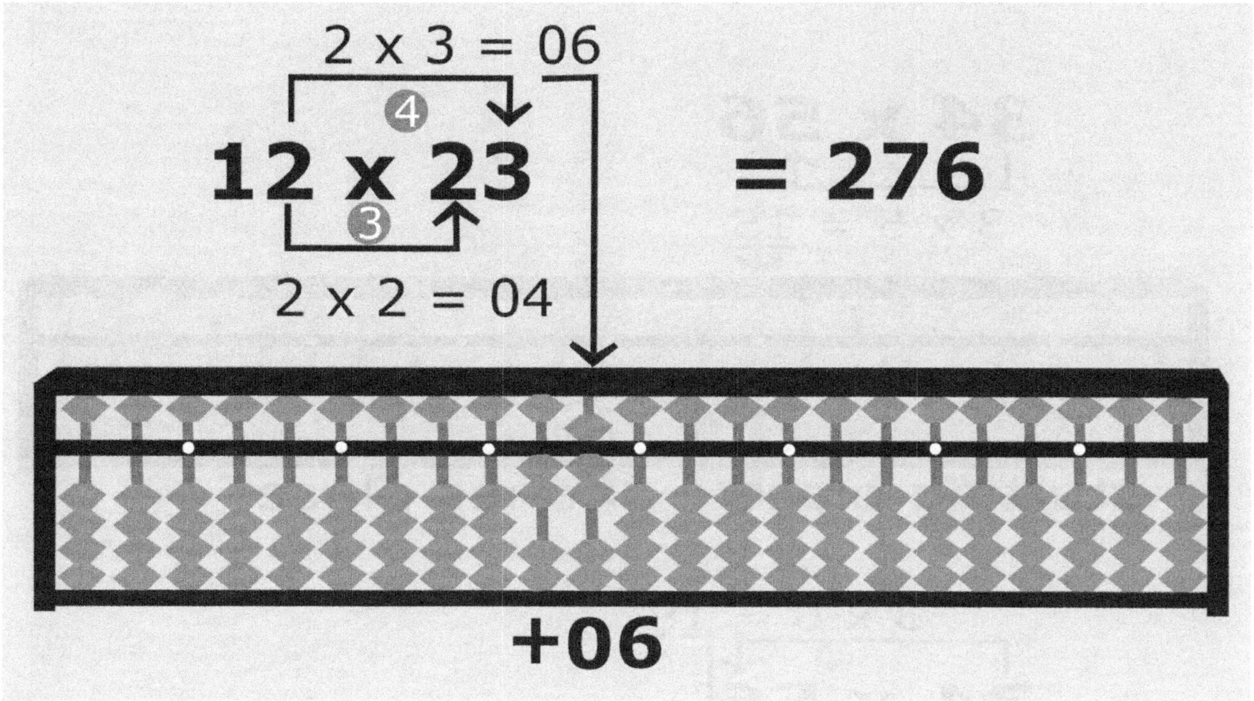

Abacus Fun Fact:

In some parts of the world, the Abacus is still used in everyday life by traders and people with visual disabilities.

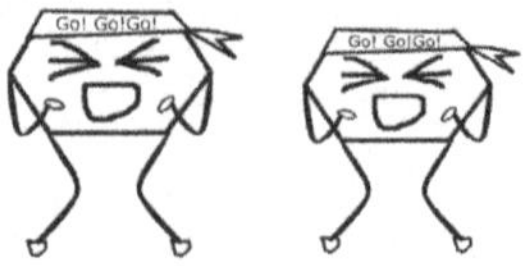
Go! Go!Go!
Go! Go!Go!

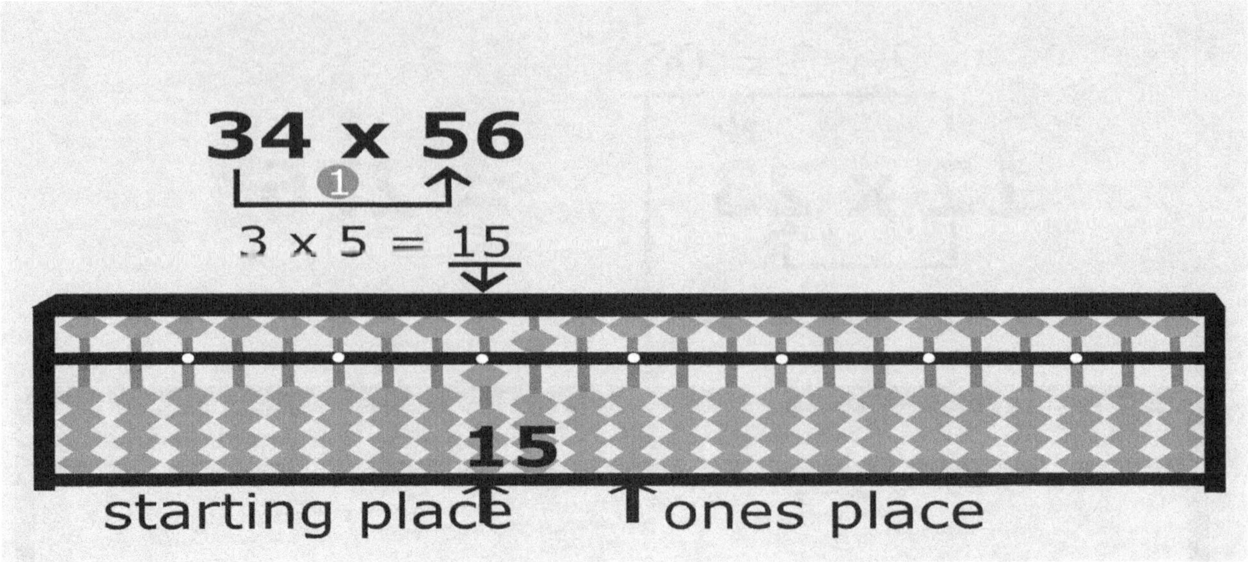
34 x 56
1
3 x 5 = 15
15
starting place
ones place

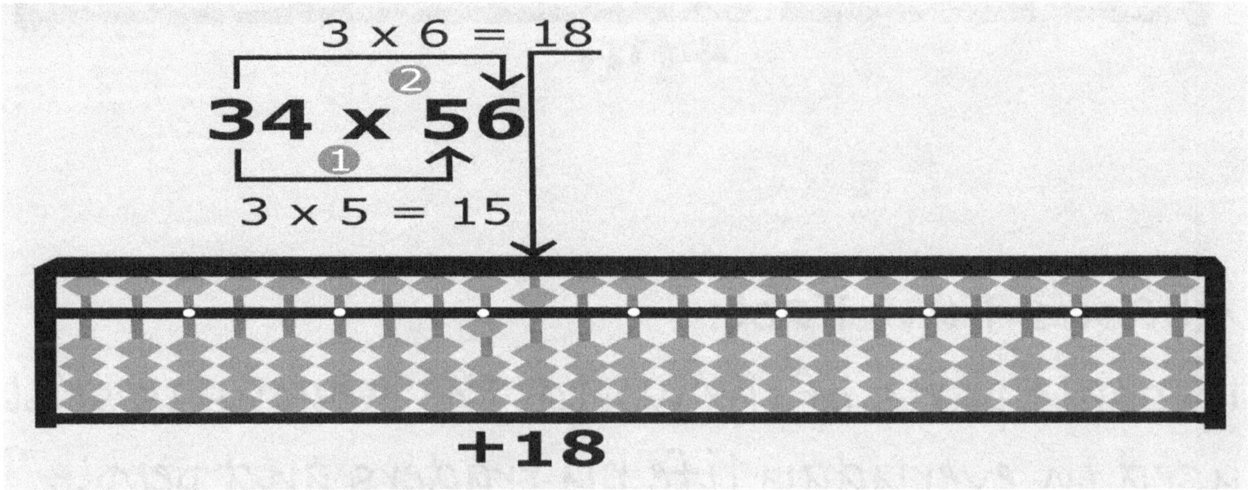
3 x 6 = 18
2
34 x 56
1
3 x 5 = 15
+18

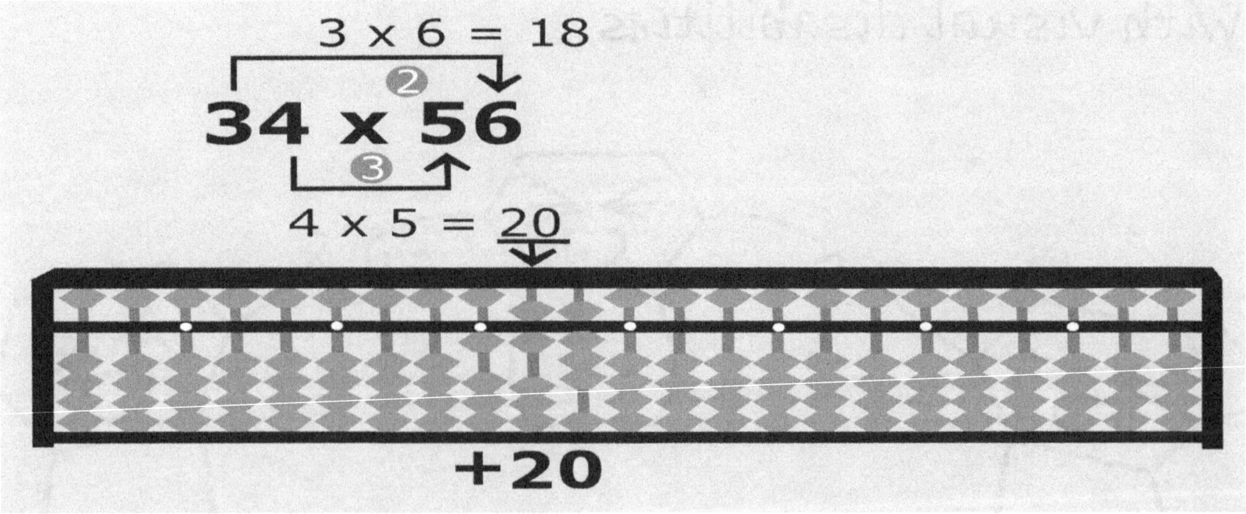
3 x 6 = 18
2
34 x 56
3
4 x 5 = 20
+20

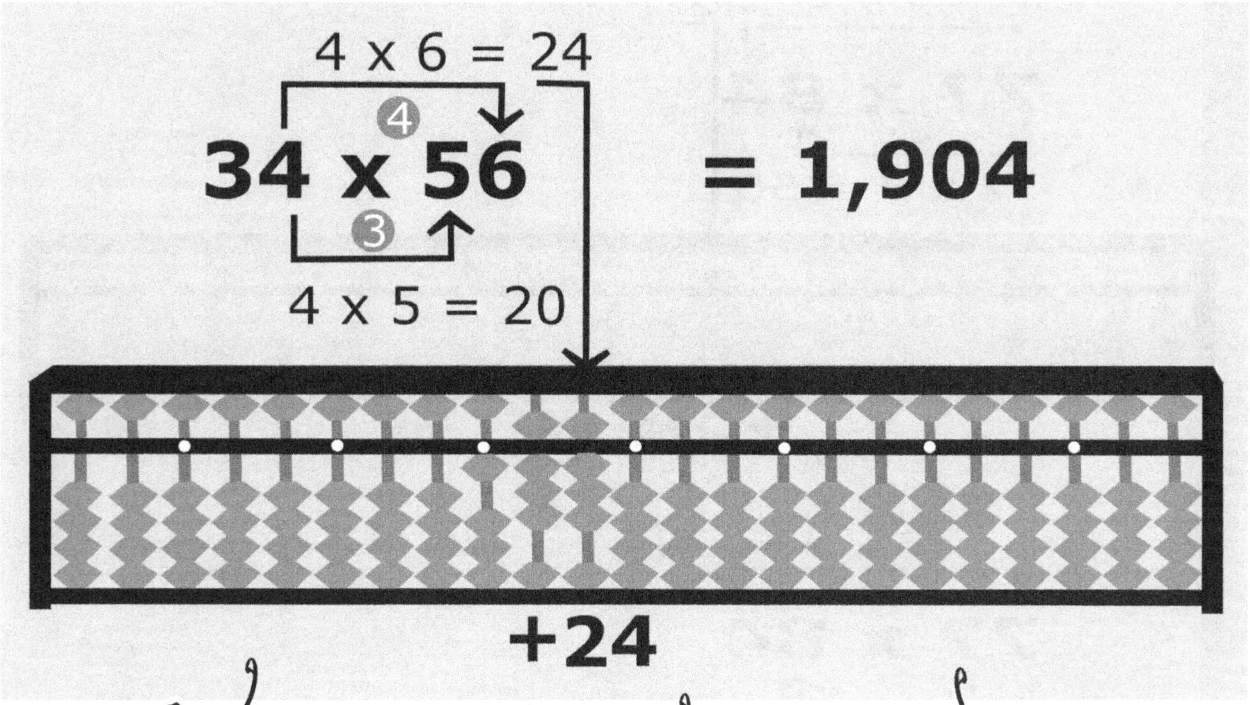
4 x 6 = 24
4
34 x 56
3
4 x 5 = 20
= 1,904
+24

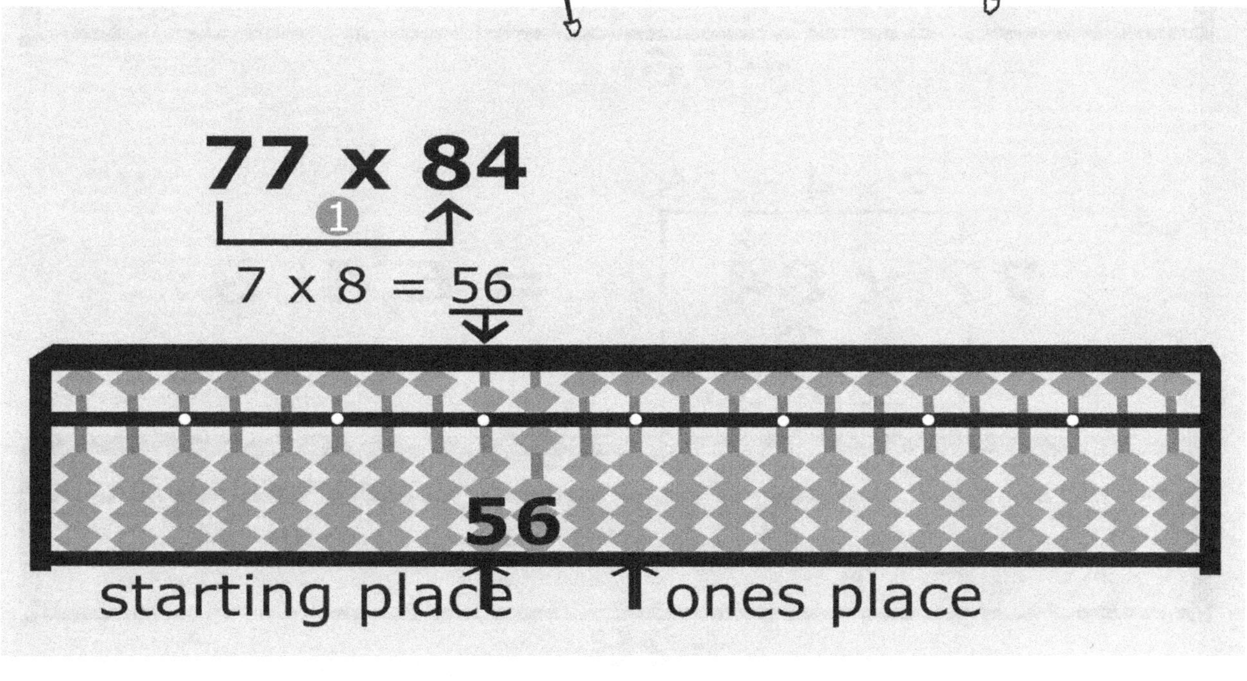
77 x 84
1
7 x 8 = 56
56
starting place
ones place

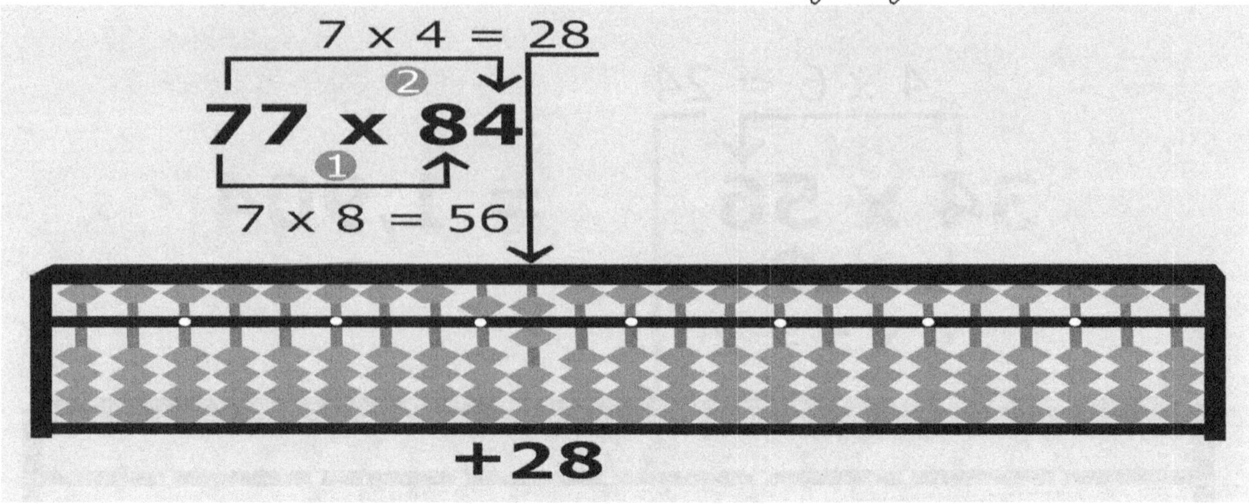
7 x 4 = 28
2
77 x 84
1
7 x 8 = 56
+28

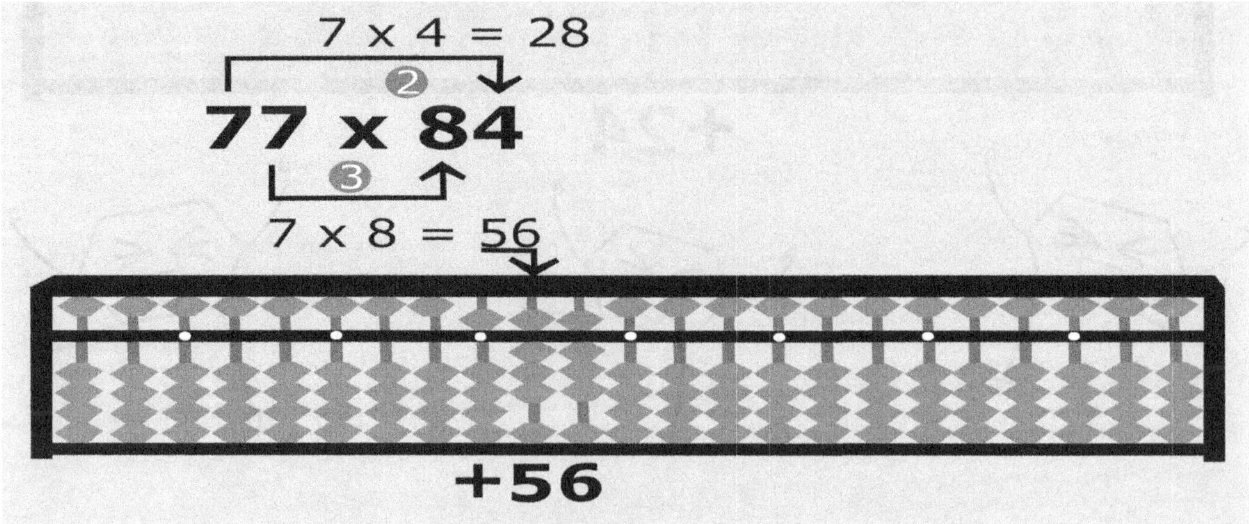
7 x 4 = 28
2
77 x 84
3
7 x 8 = 56
+56

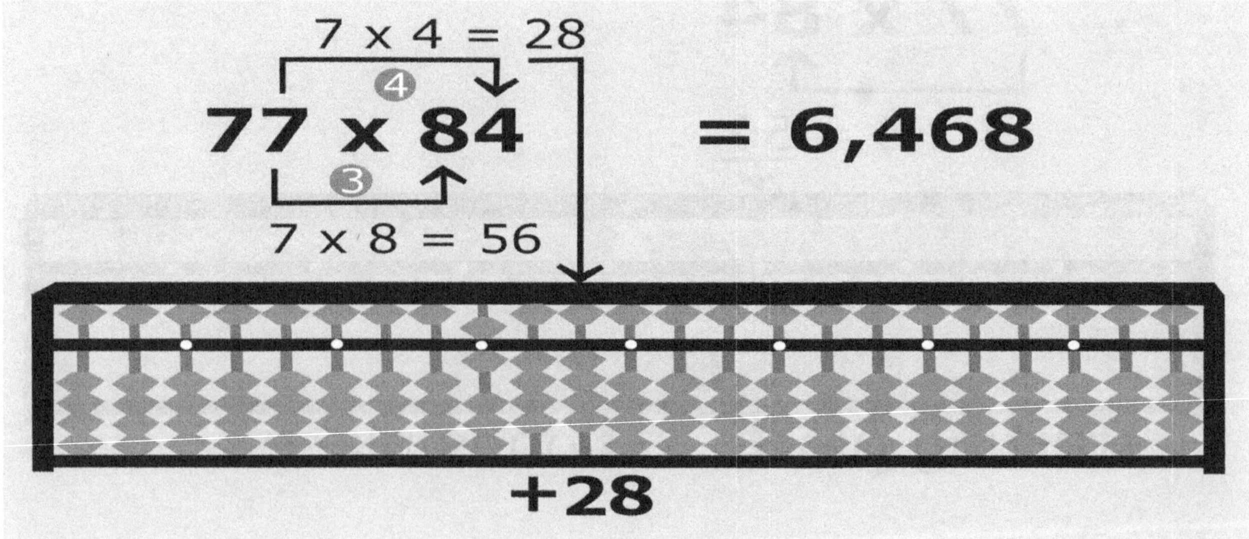
7 x 4 = 28
4
77 x 84
3
7 x 8 = 56
= 6,468
+28

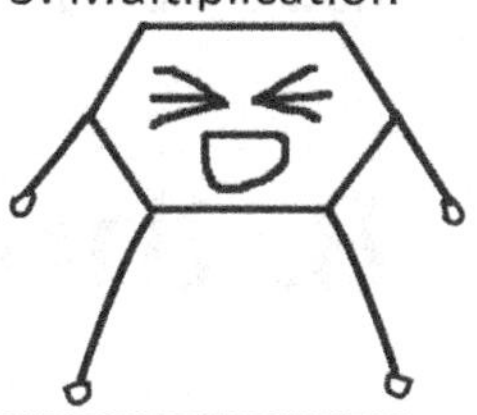

Practice for 2 digits times 2 digits

1) 46 X 41 =	15) 71 X 49 =
2) 30 X 19 =	16) 56 X 91 =
3) 13 X 48 =	17) 65 X 60 =
4) 79 X 17 =	18) 22 X 94 =
5) 96 X 75 =	19) 88 X 13 =
6) 15 X 73 =	20) 92 X 68 =
7) 36 X 87 =	21) 69 X 16 =
8) 72 X 12 =	22) 66 X 54 =
9) 58 X 82 =	23) 41 X 96 =
10) 95 X 45 =	24) 75 X 29 =
11) 43 X 56 =	25) 16 X 18 =
12) 99 X 84 =	26) 64 X 74 =
13) 38 X 13 =	27) 77 X 61 =
14) 65 X 25 =	28) 53 X 29 =

8. three digits times two digits

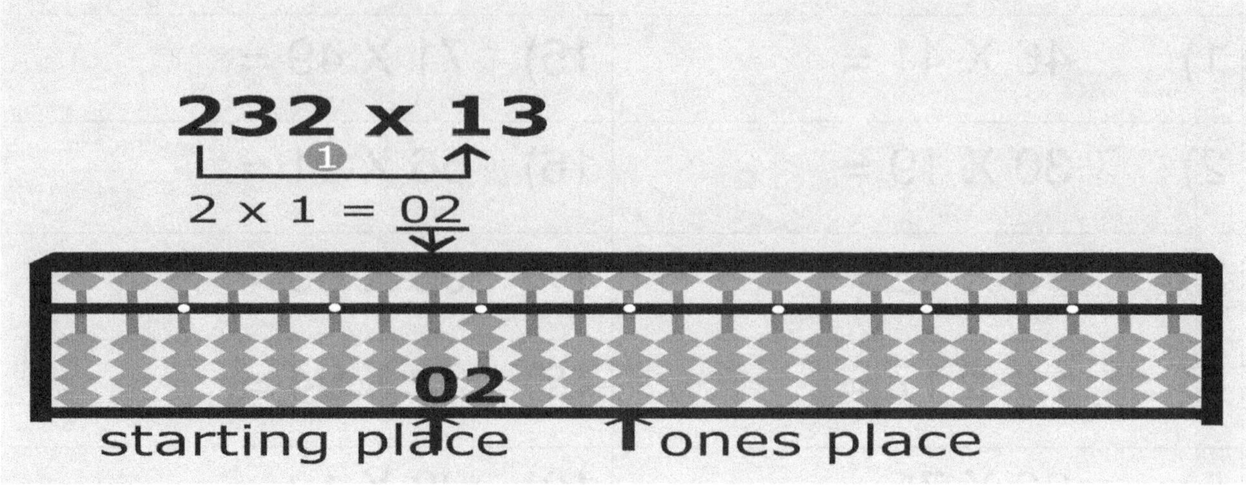

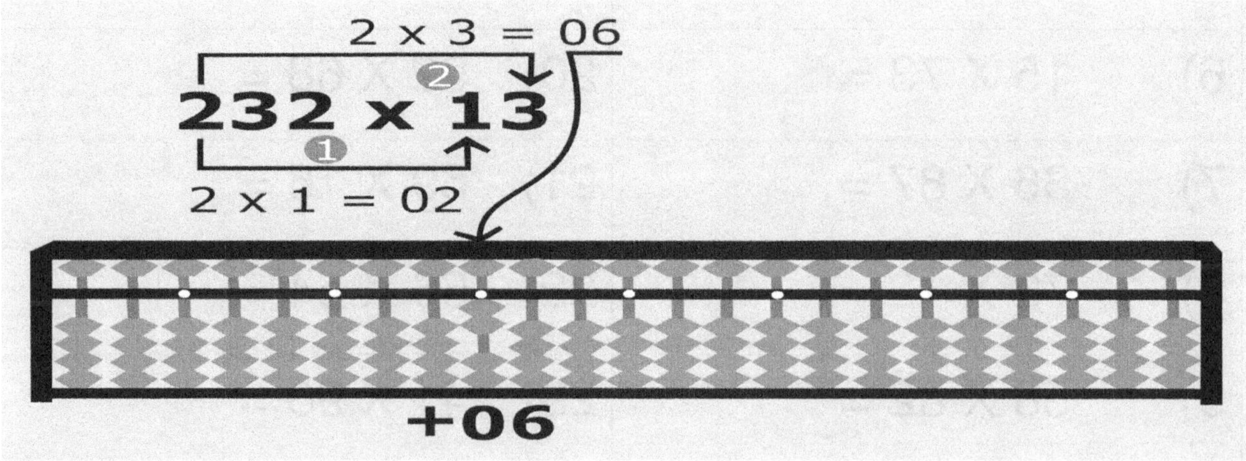

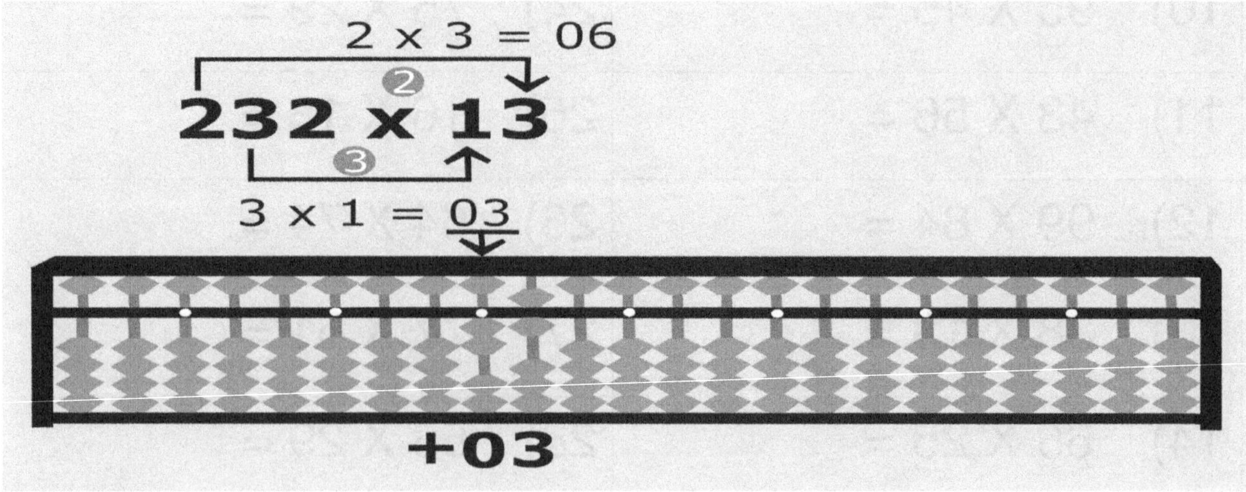

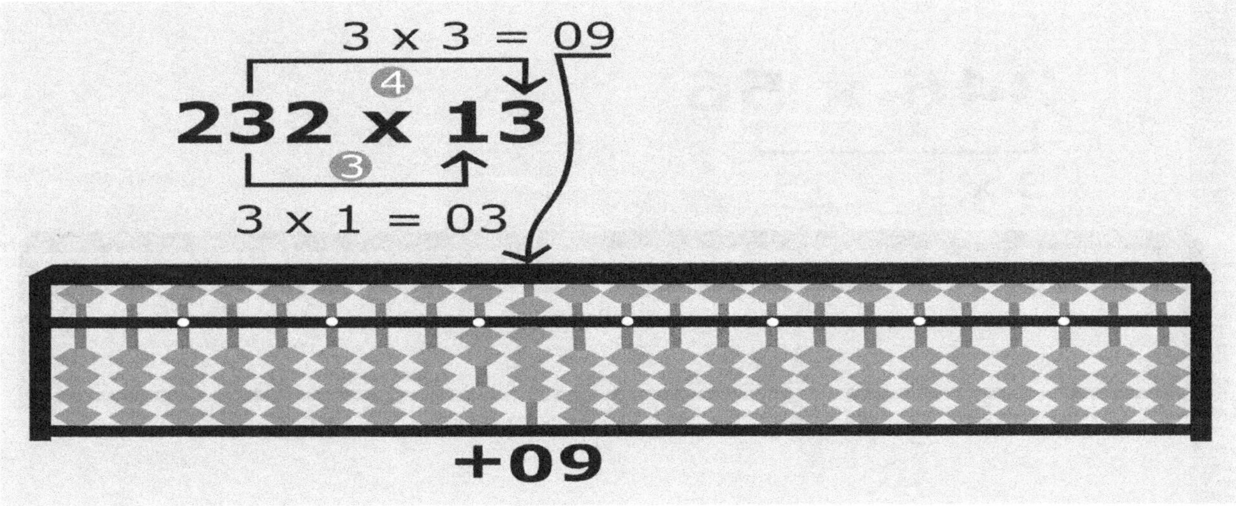
3 x 3 = 09
4
232 x 13
3
3 x 1 = 03
+09

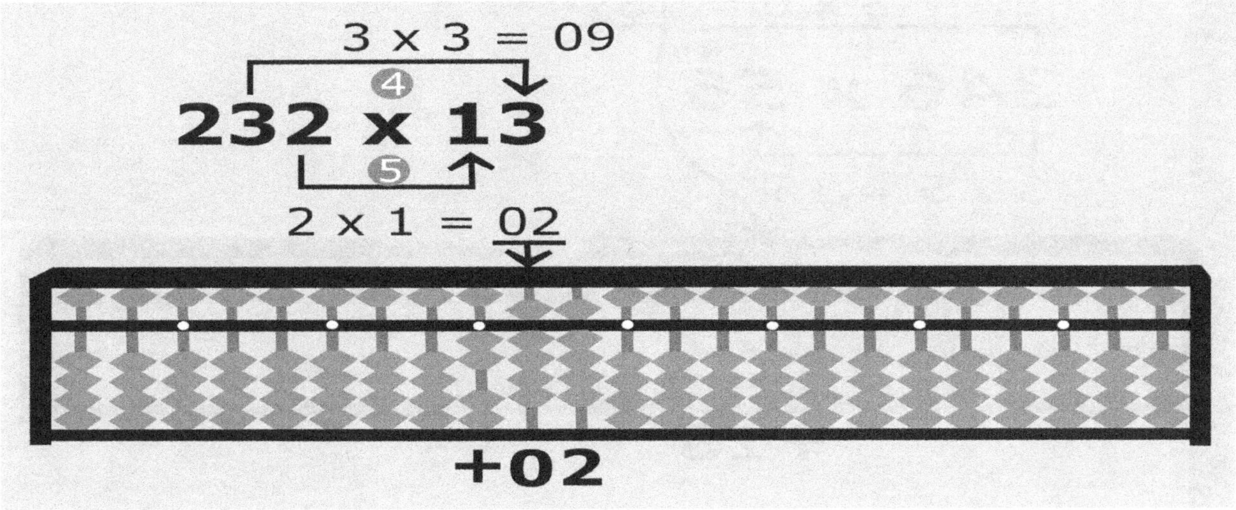
3 x 3 = 09
4
232 x 13
5
2 x 1 = 02
+02

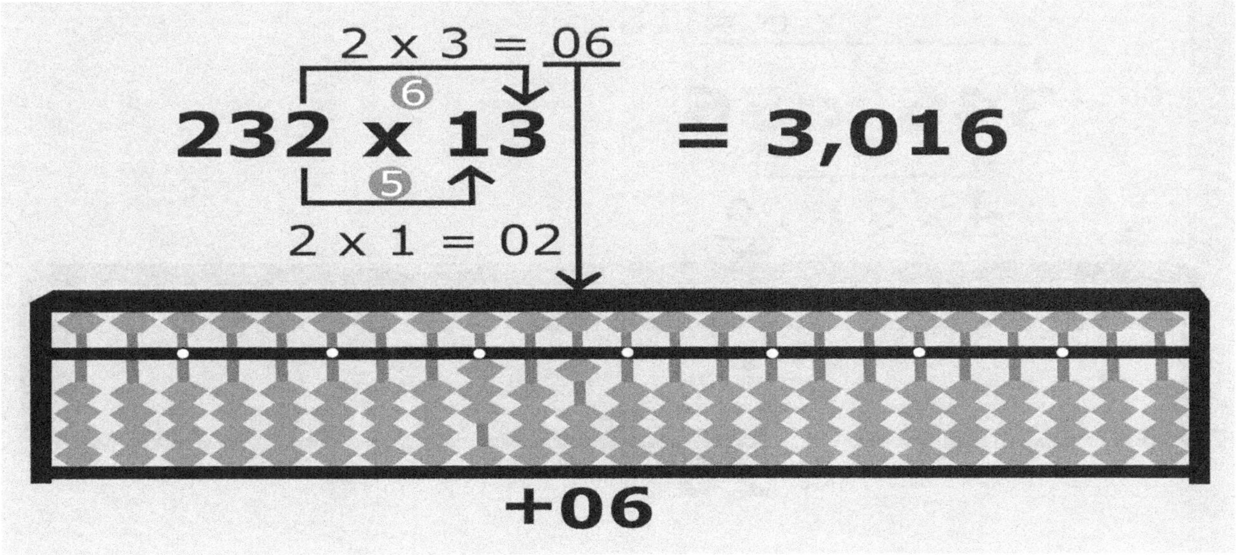
2 x 3 = 06
6
232 x 13
= 3,016
5
2 x 1 = 02
+06

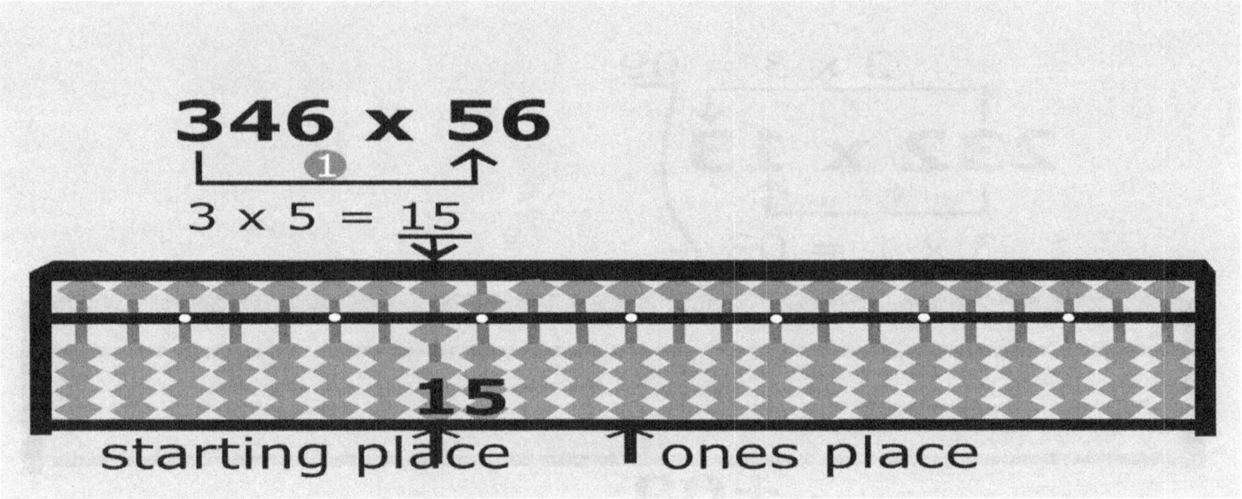
346 x 56
1
3 x 5 = 15
15
starting place
ones place

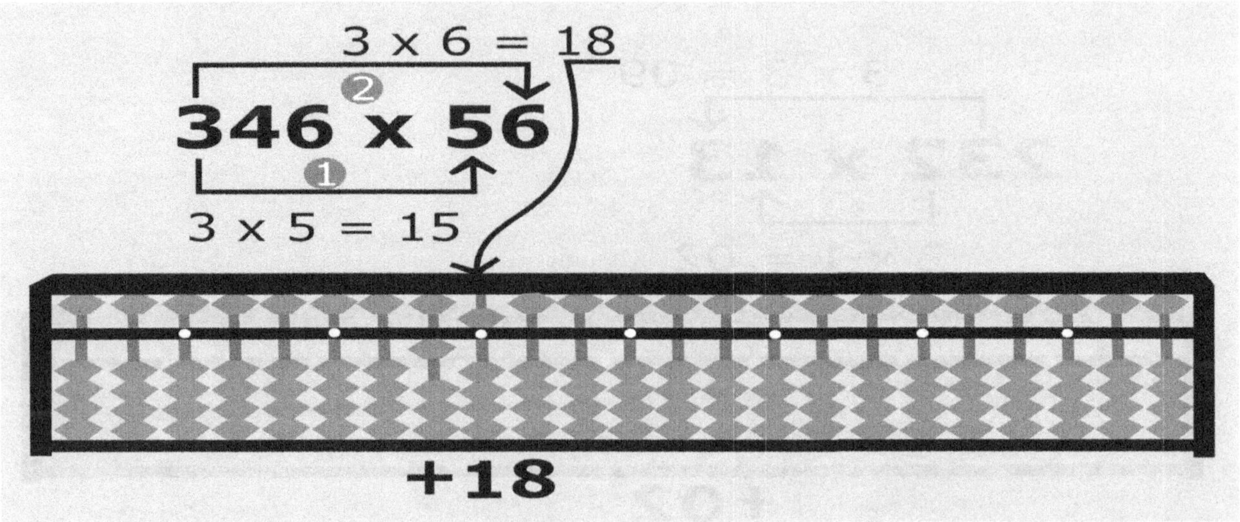
3 x 6 = 18
2
346 x 56
1
3 x 5 = 15
+18

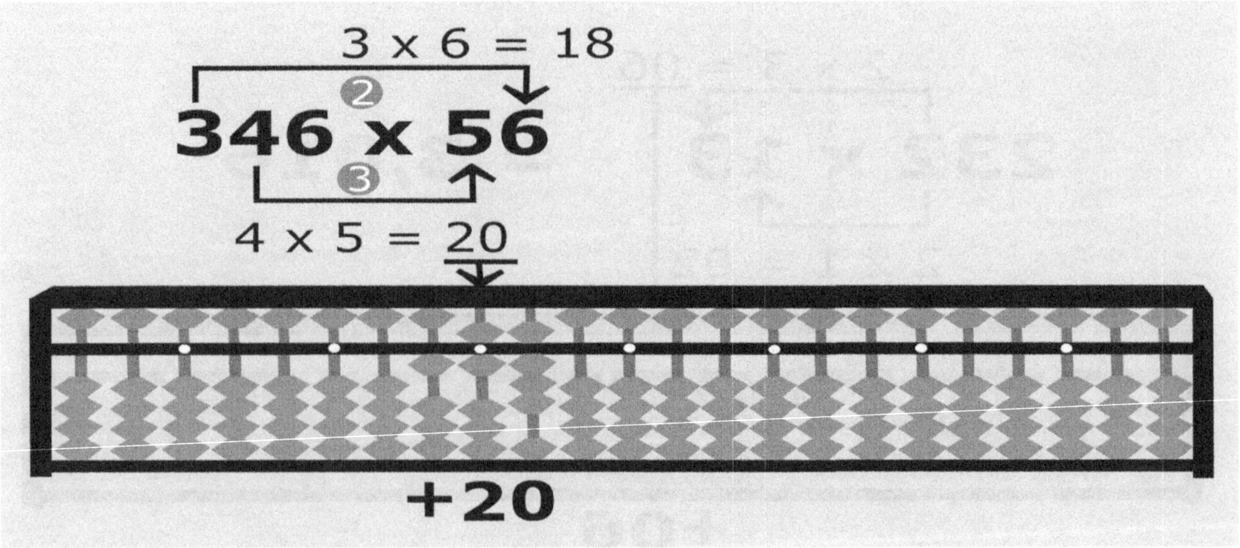
3 x 6 = 18
2
346 x 56
3
4 x 5 = 20
+20

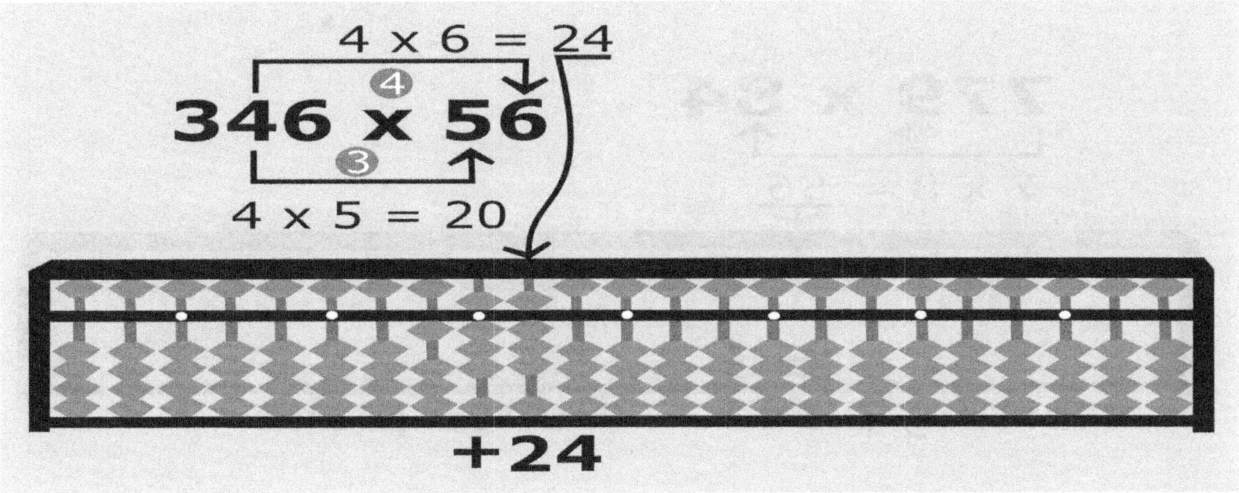
4 x 6 = 24
4
346 x 56
3
4 x 5 = 20
+24

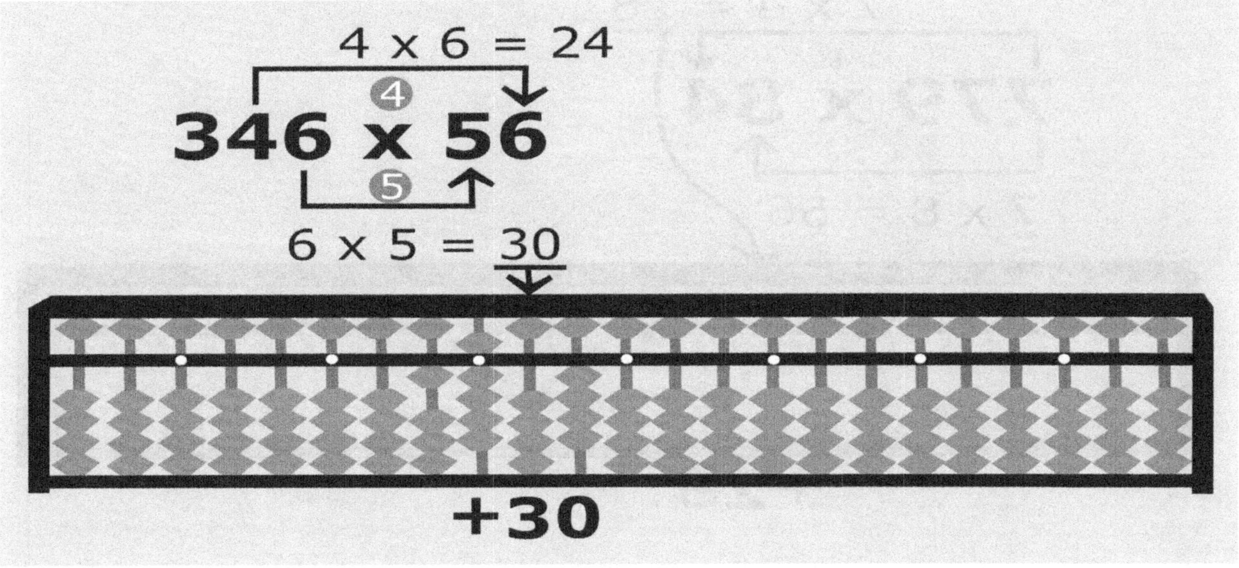
4 x 6 = 24
4
346 x 56
5
6 x 5 = 30
+30

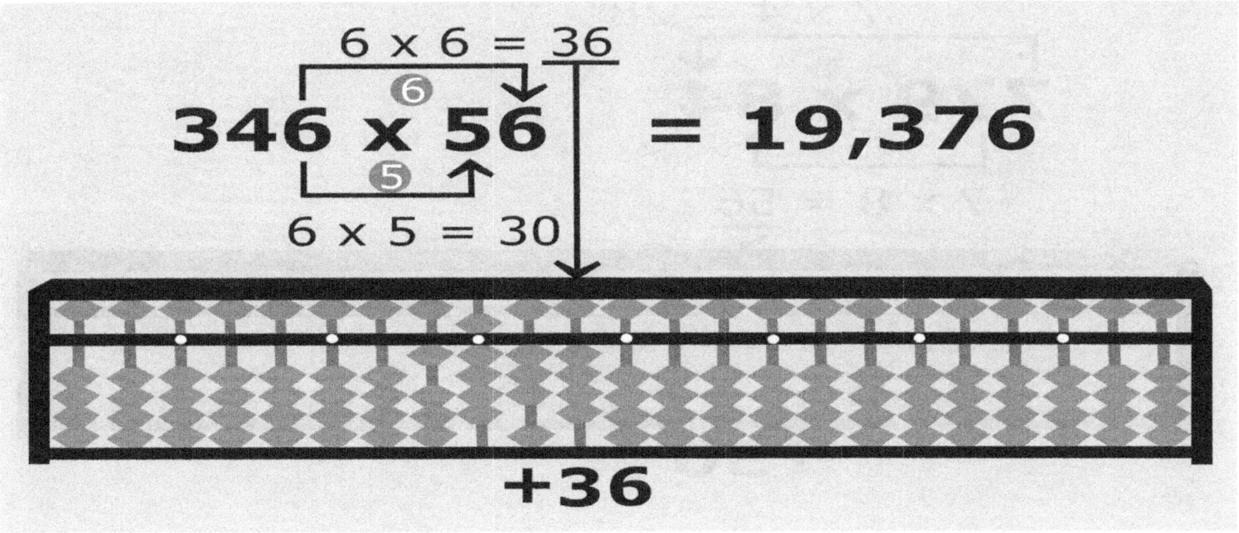
6 x 6 = 36
6
346 x 56
= 19,376
5
6 x 5 = 30
+36

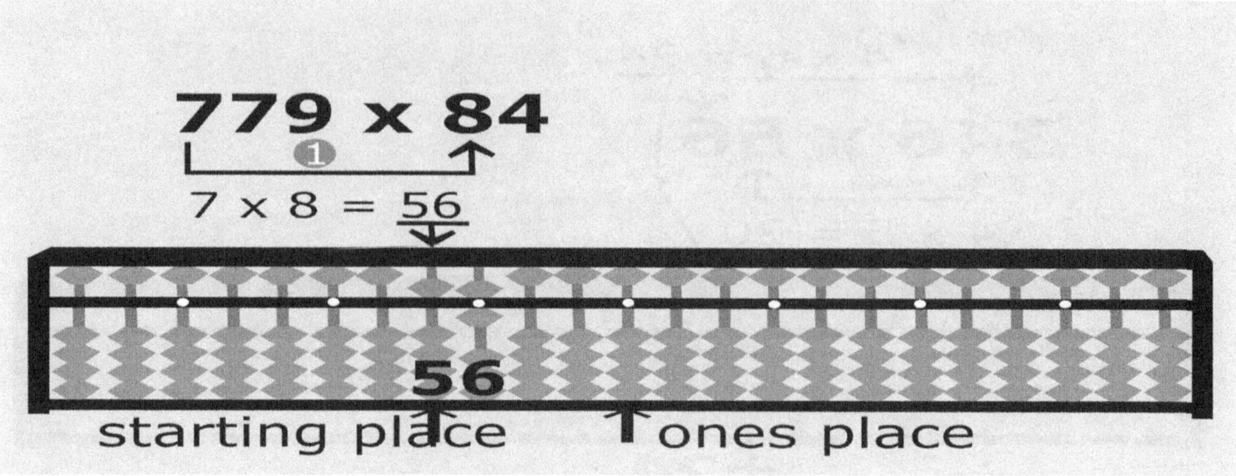
779 x 84
1
7 x 8 = 56
56
starting place
ones place

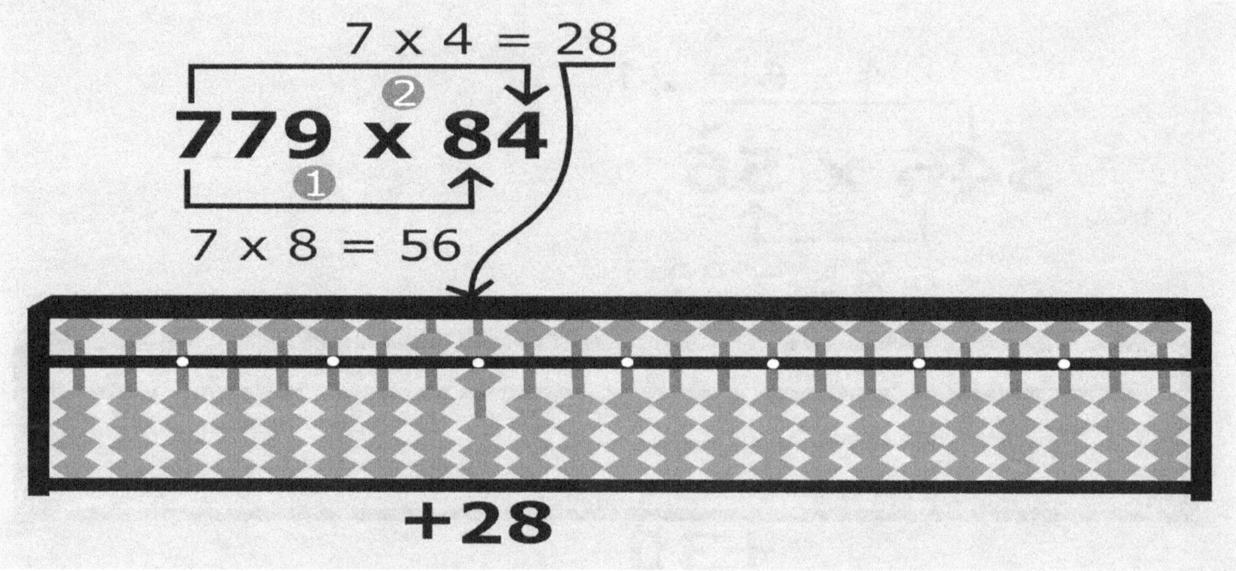
7 x 4 = 28
2
779 x 84
1
7 x 8 = 56
+28

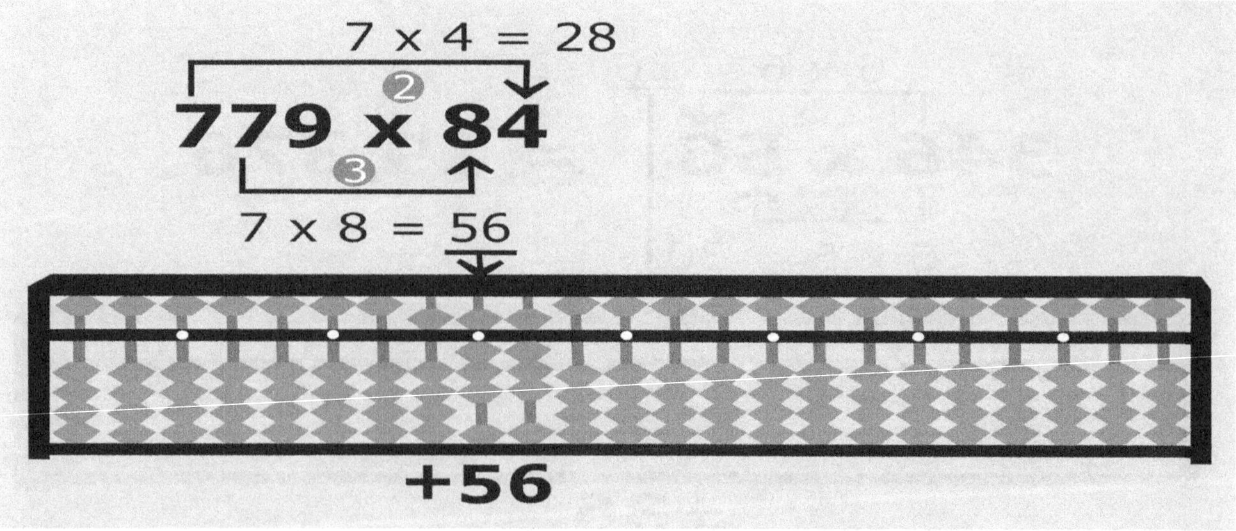
7 x 4 = 28
2
779 x 84
3
7 x 8 = 56
+56

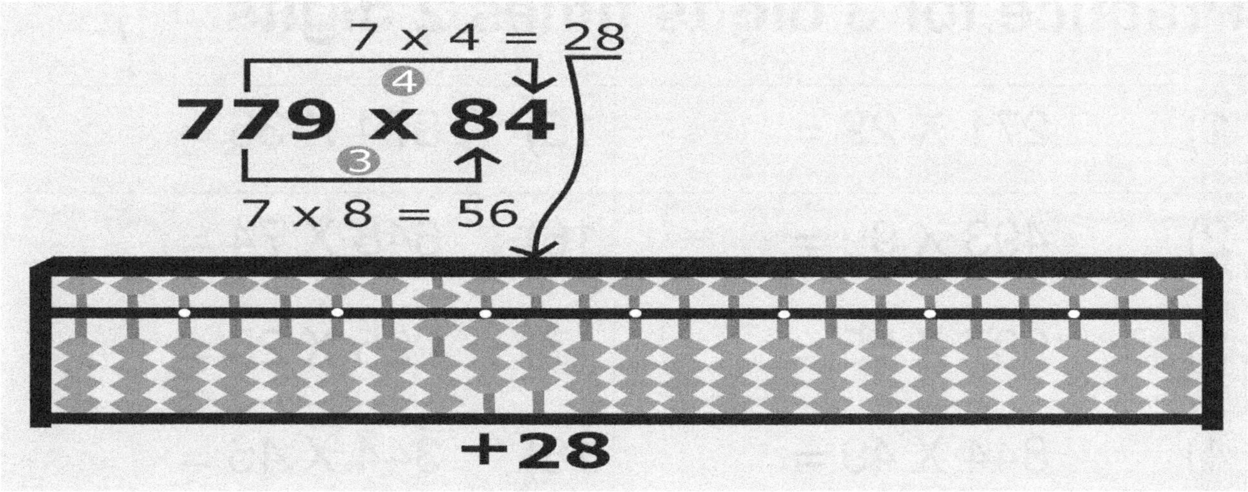
7 x 4 = 28
4
779 x 84
3
7 x 8 = 56
+28

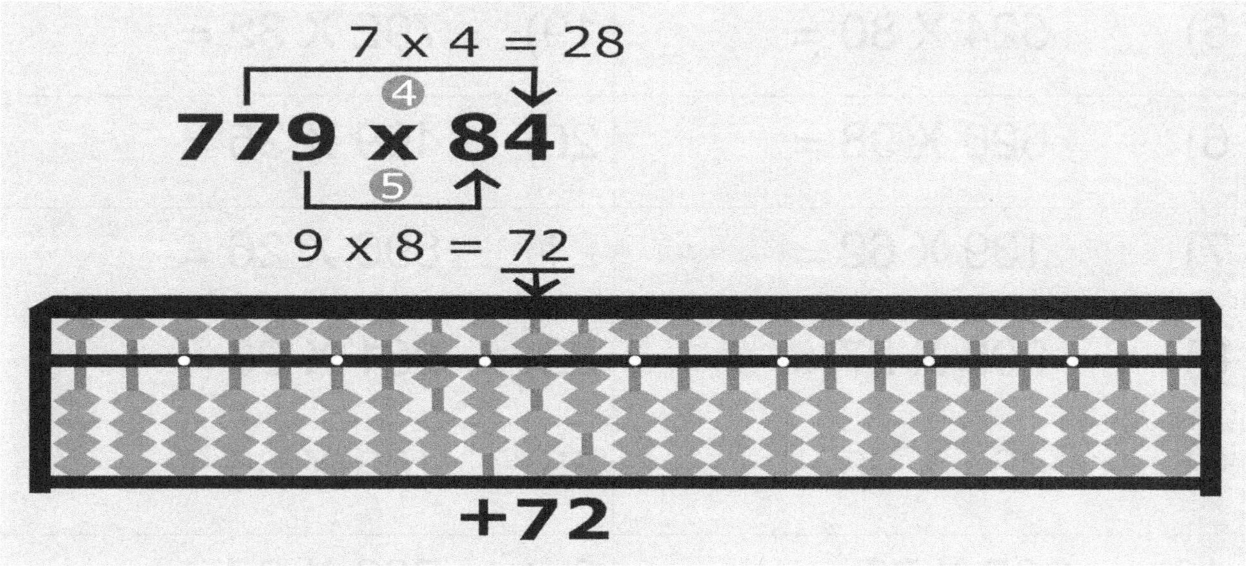
7 x 4 = 28
4
779 x 84
5
9 x 8 = 72
+72

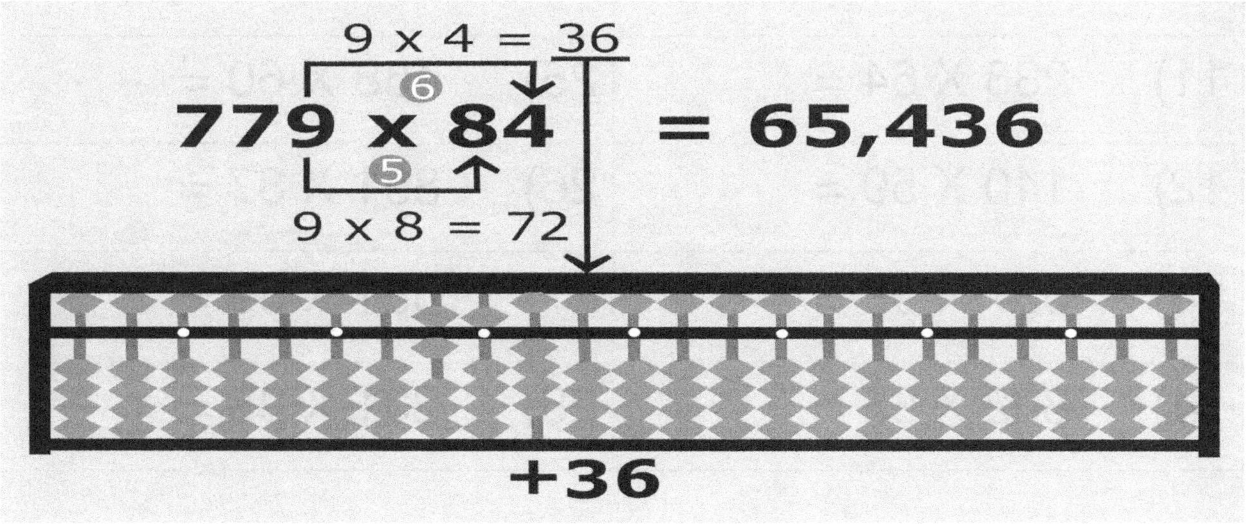
9 x 4 = 36
6
779 x 84
= 65,436
5
9 x 8 = 72
+36

Practice for 3 digits times 2 digits

1) 271 X 22 =	15) 874 X 38 =
2) 493 X 99 =	16) 548 X 74 =
3) 997 X 55 =	17) 225 X 25 =
4) 844 X 49 =	18) 344 X 46 =
5) 524 X 80 =	19) 792 X 32 =
6) 626 X 98 =	20) 499 X 35 =
7) 139 X 62 =	21) 896 X 26 =
8) 398 X 27 =	22) 281 X 28 =
9) 635 X 28 =	23) 103 X 65 =
10) 965 X 73 =	24) 769 X 97 =
11) 238 X 64 =	25) 468 X 60 =
12) 140 X 59 =	26) 851 X 87 =
13) 337 X 98 =	27) 551 X 89 =
14) 802 X 63 =	28) 201 X 38 =

Abacus Fun Fact:
The Abacus have been used by many civilizations including China, Greece, Rome, Russia, Japan and more.

9. Multiplication with decimal

The multiplication algorithm is the same for numbers containing decimals. However, it is important to distinguish the location of the decimal point.

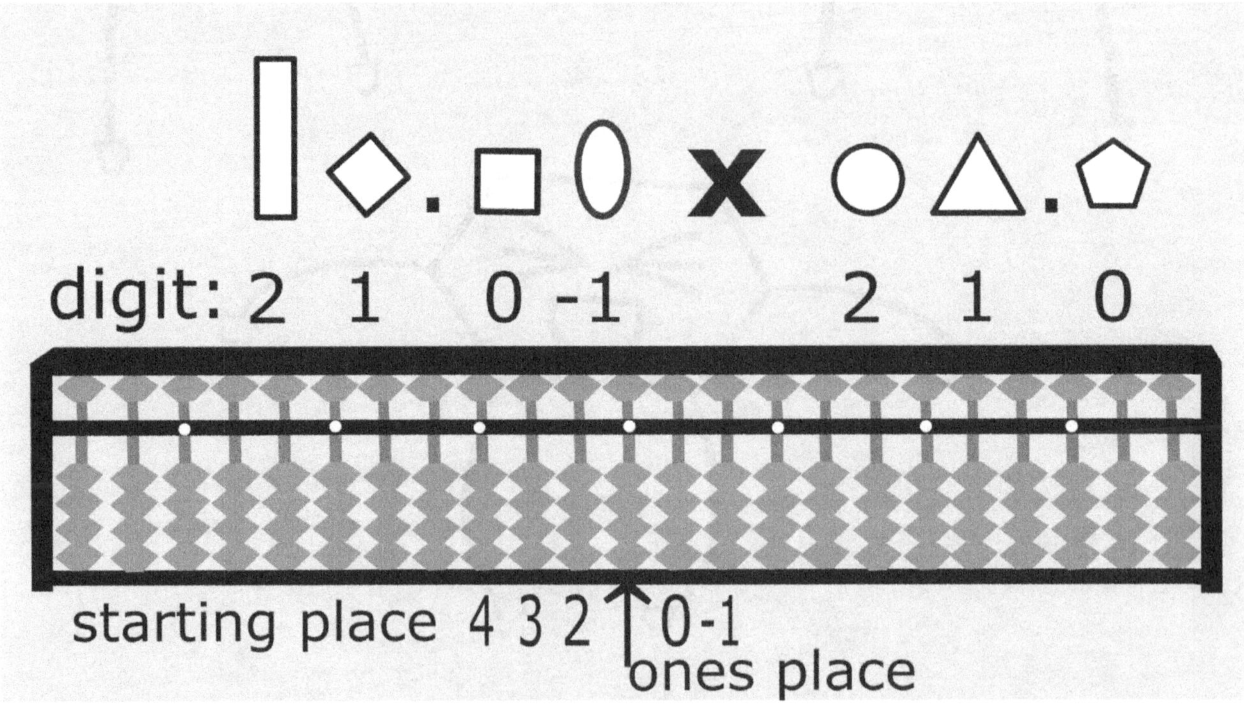

The diagram above shows the number with 2 decimals, ▯◇.□⬯ , multiplied by another number with 1 decimal, ○△.⬠. The final product should have 3 decimals (2 decimals <□, ⬯> + 1 decimal <⬠> = 3 decimals).

Integers for the starting place

Only the integer's digit will be added together to identify the starting place for the product on the abacus.

In the diagram of ▯◇.□0 x ○△.⬠, ▯, ◇, ○ and △ are integers; therefore, the starting place for this problem would be in the 4th place on the abacus (2 digits <▯, ◇> + 2 digits < ○, △ = 4 digits).

Follow multiplication sequence

The multiplication algorithm goes in order so that if the starting place for the first number has been distinguished, then the starting places for the rest of the numbers would be easy to identify. If the starting place for ▯ x ○ is in the 4th place on the abacus, the starting place for ▯ x △ would be in the 3rd place on the abacus etc.

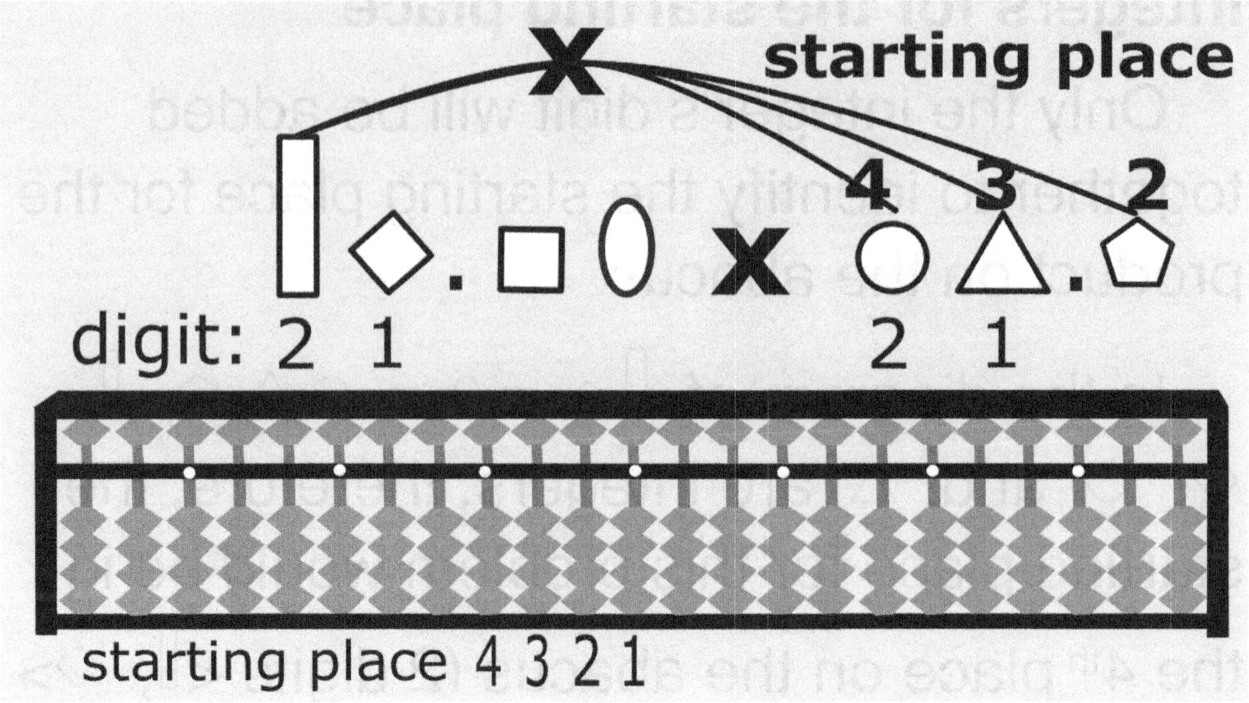
starting place
4 3 2
digit: 2 1 2 1
starting place 4 3 2 1

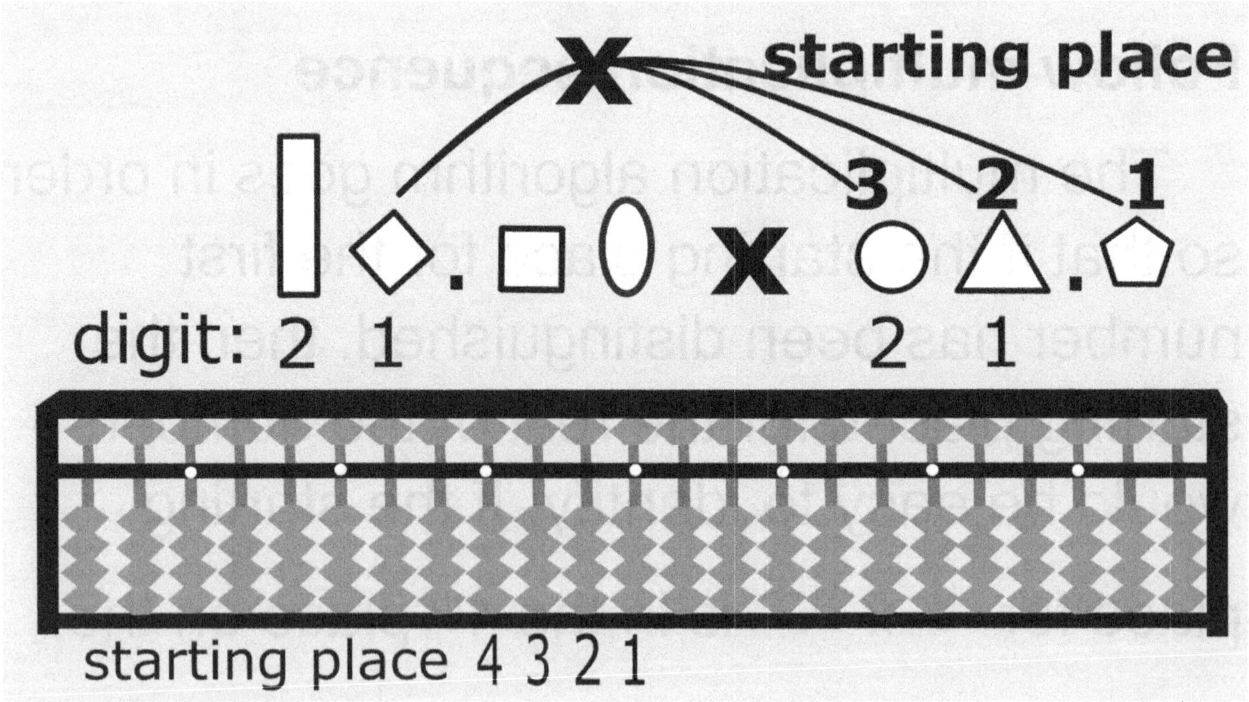
starting place
3 2 1
digit: 2 1 2 1
starting place 4 3 2 1

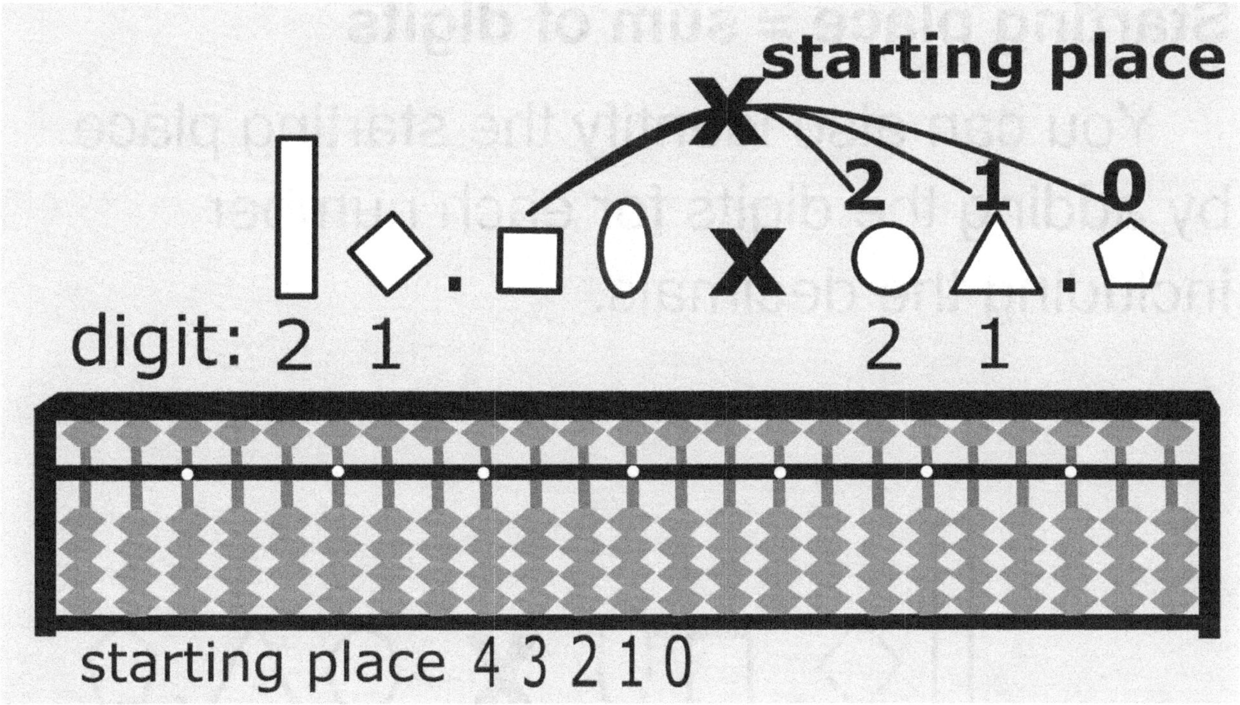
starting place
2
1
0
digit: 2 1
2 1
starting place 4 3 2 1 0

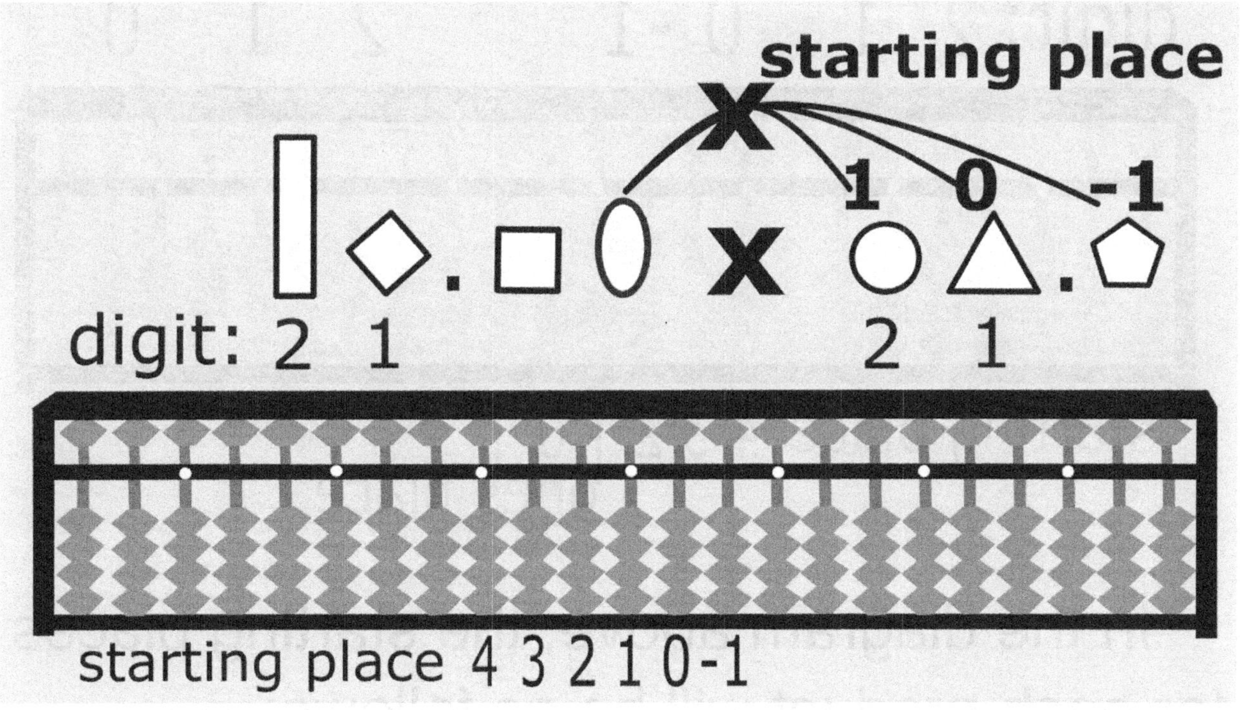
starting place
1
0
-1
digit: 2 1
2 1
starting place 4 3 2 1 0-1

Starting place = sum of digits

You can also identify the starting place by adding the digits for each number including the decimals.

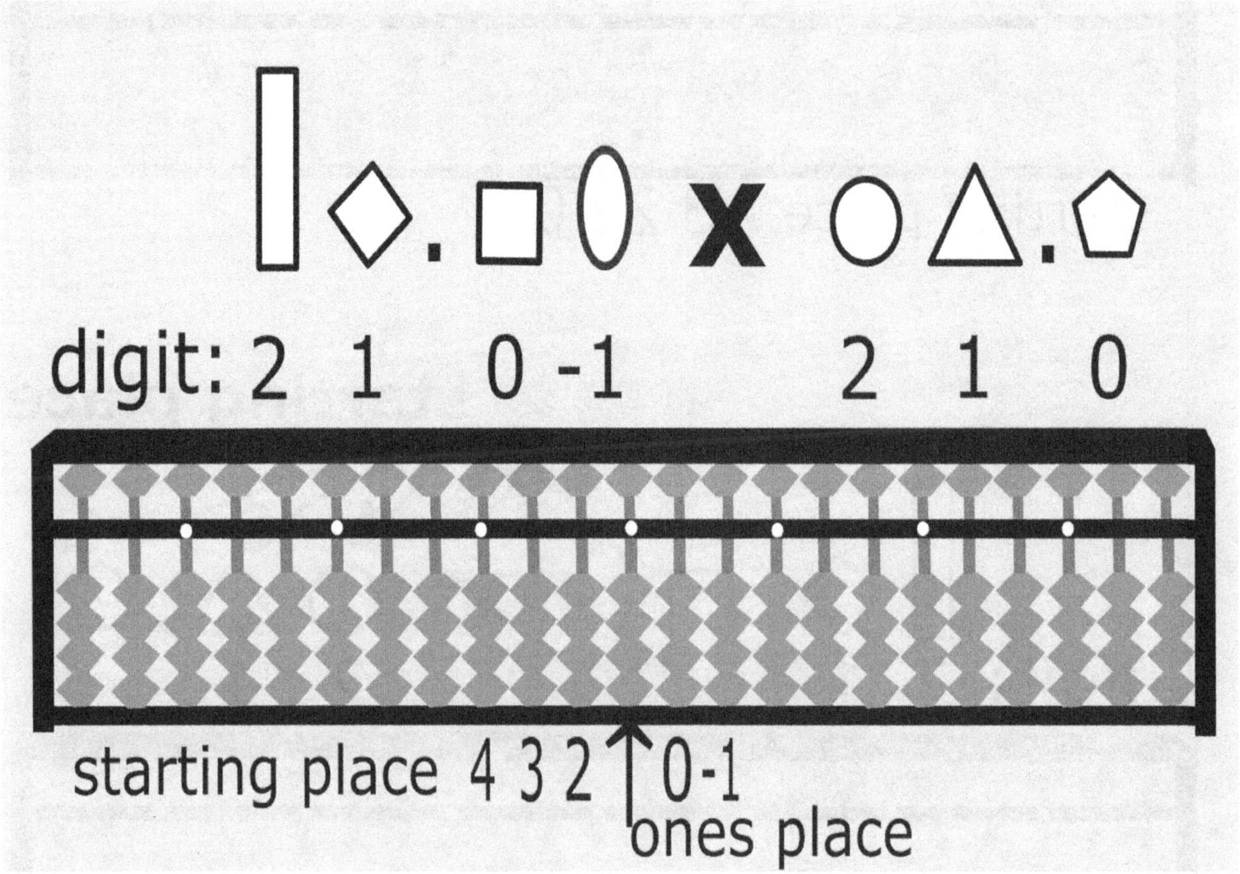

In the diagram above, the starting places for each product will be as follows:

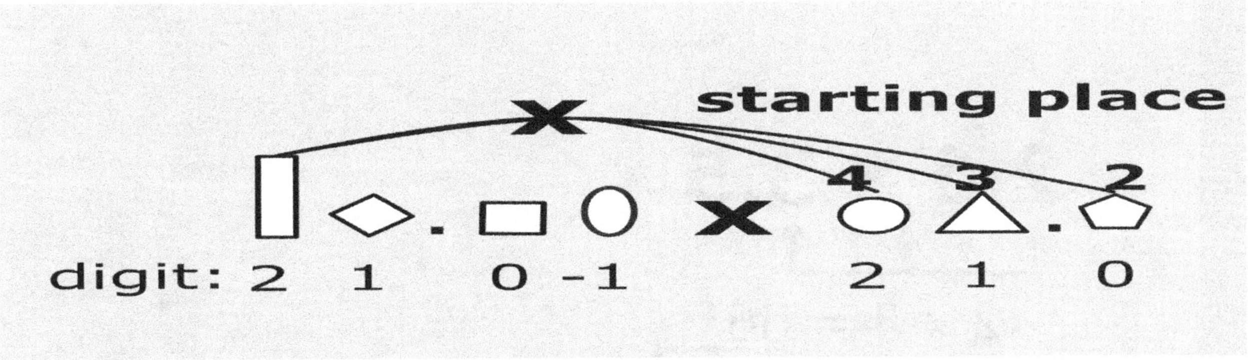
starting place
4 3 2
digit: 2 1 0 -1 2 1 0

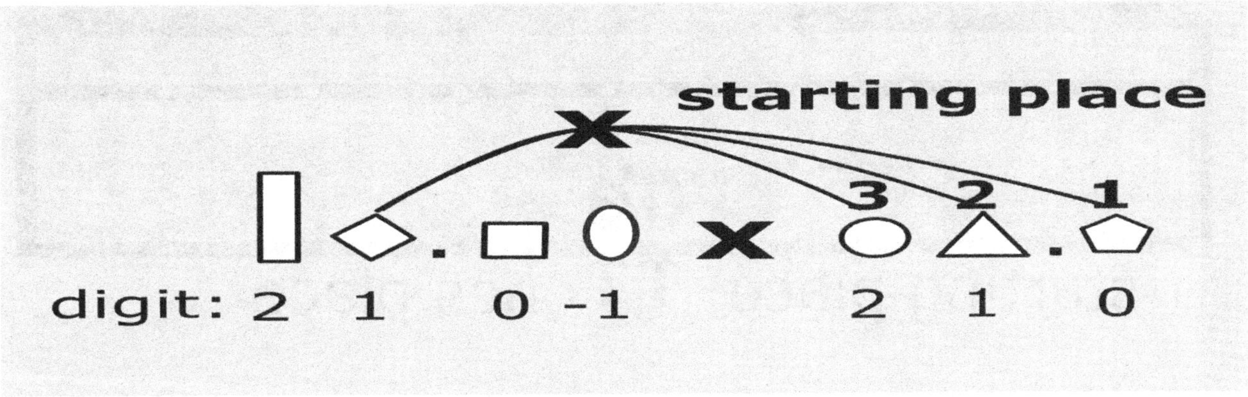
starting place
3 2 1
digit: 2 1 0 -1 2 1 0

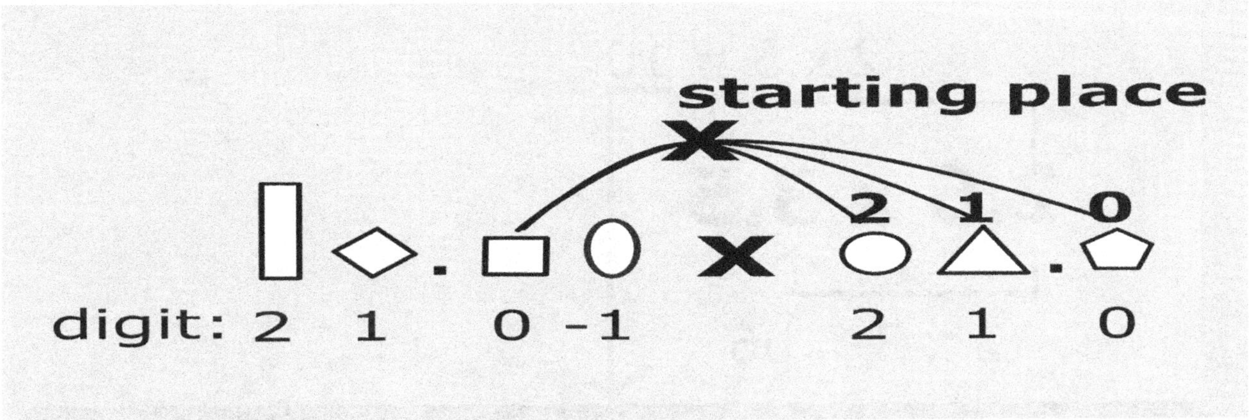
starting place
2 1 0
digit: 2 1 0 -1 2 1 0

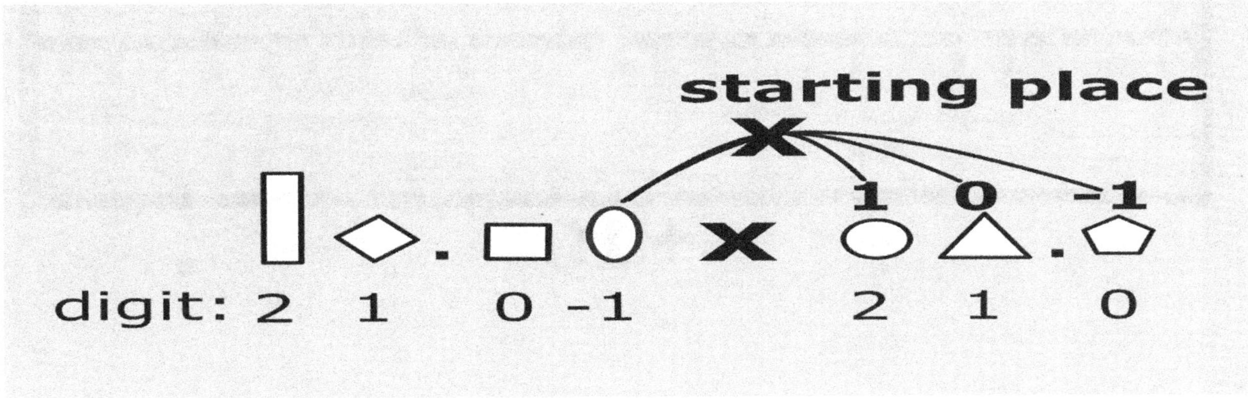
starting place
1 0 -1
digit: 2 1 0 -1 2 1 0

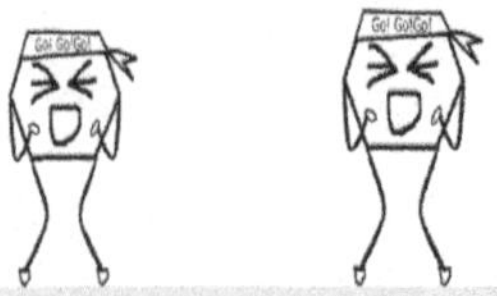

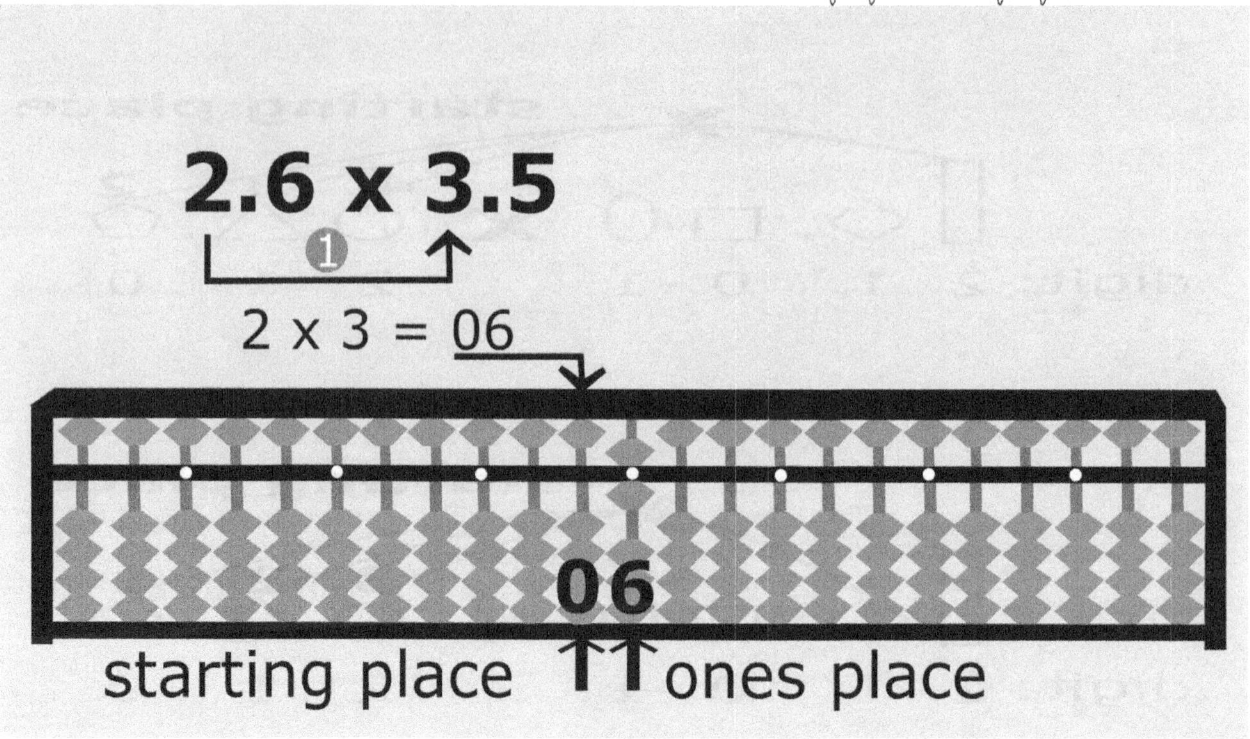
2.6 x 3.5
1
2 x 3 = 06
06
starting place
ones place

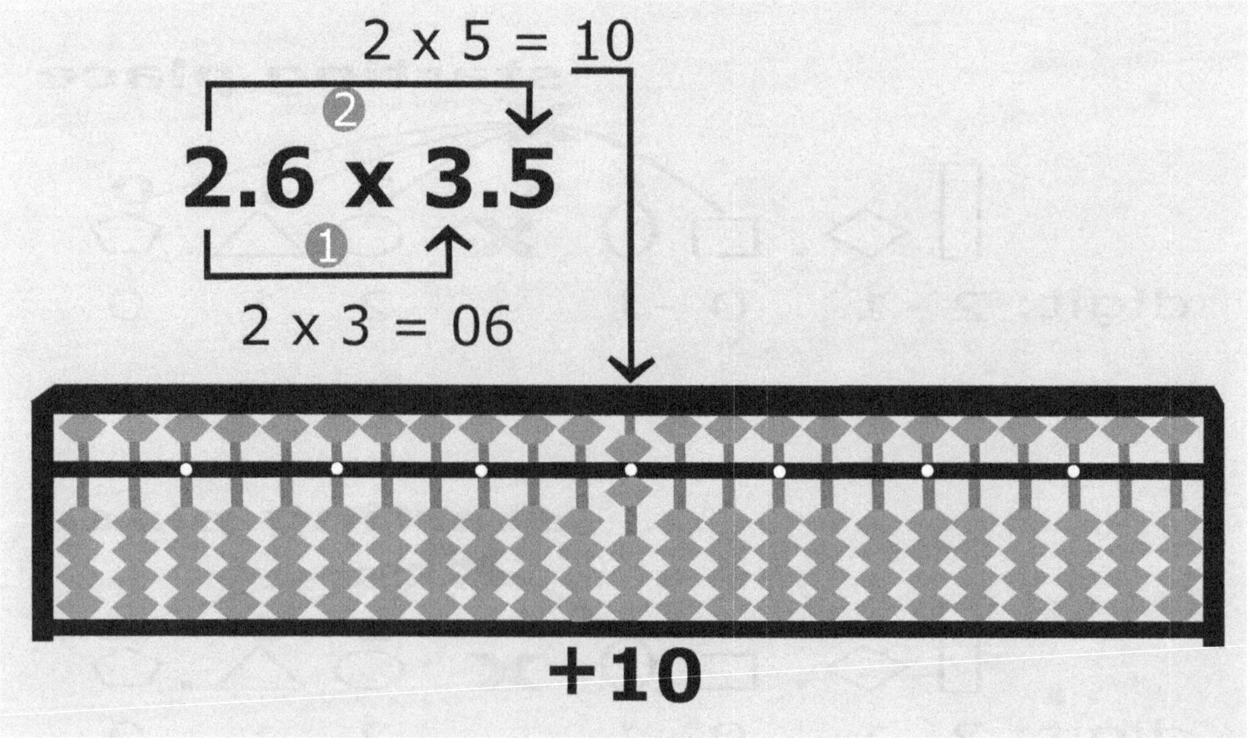
2 x 5 = 10
2
2.6 x 3.5
1
2 x 3 = 06
+10

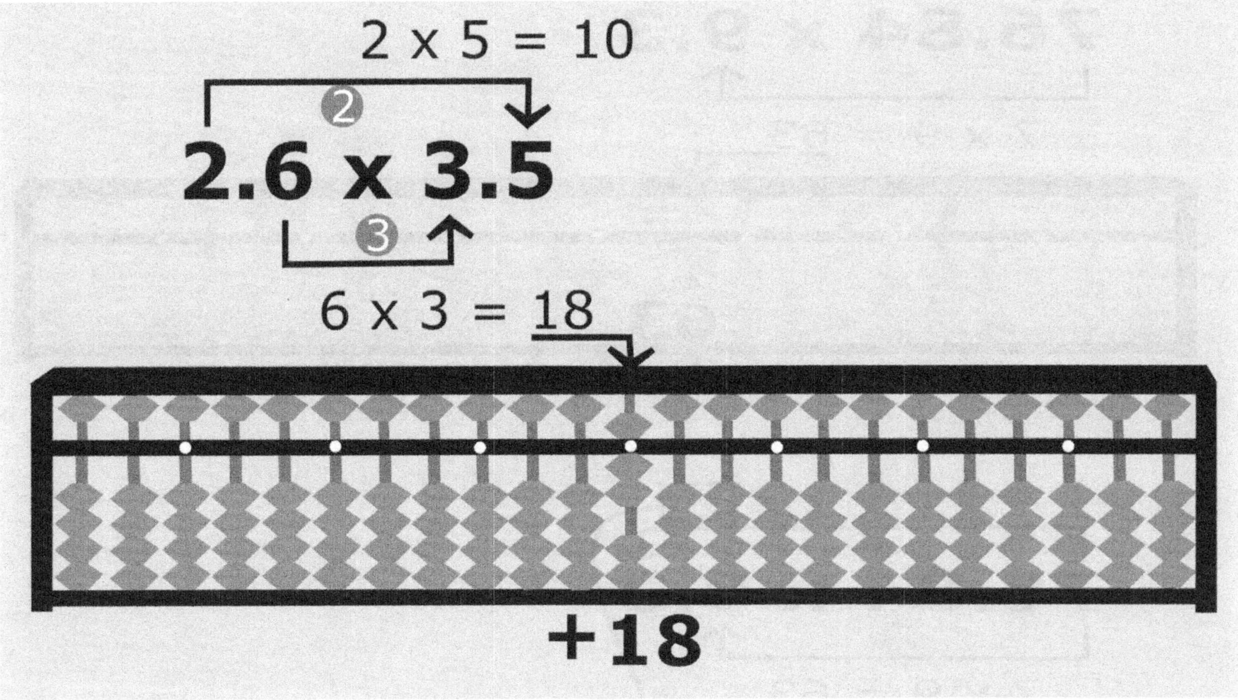
2 x 5 = 10
2
2.6 x 3.5
3
6 x 3 = 18
+18

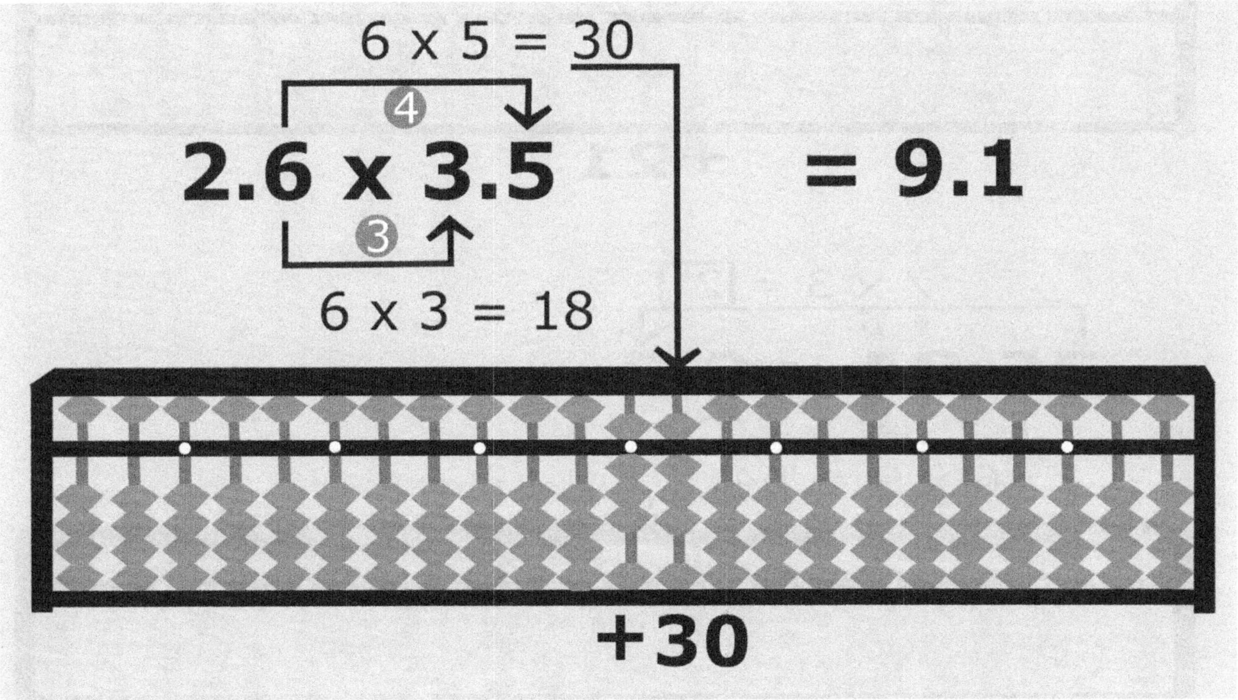
6 x 5 = 30
4
2.6 x 3.5
= 9.1
3
6 x 3 = 18
+30

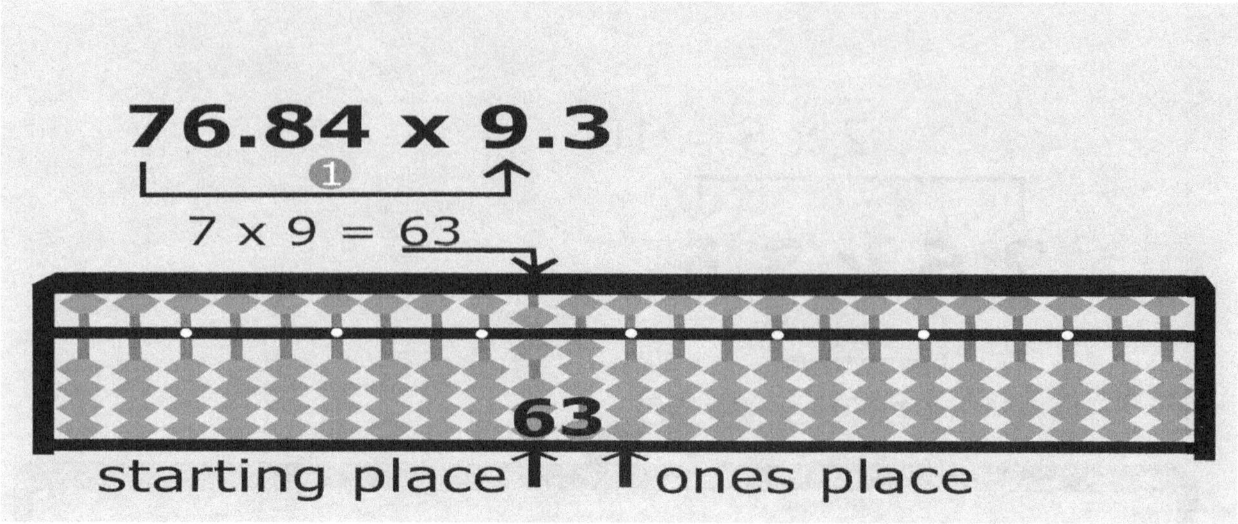
76.84 x 9.3
1
7 x 9 = 63
63
starting place
ones place

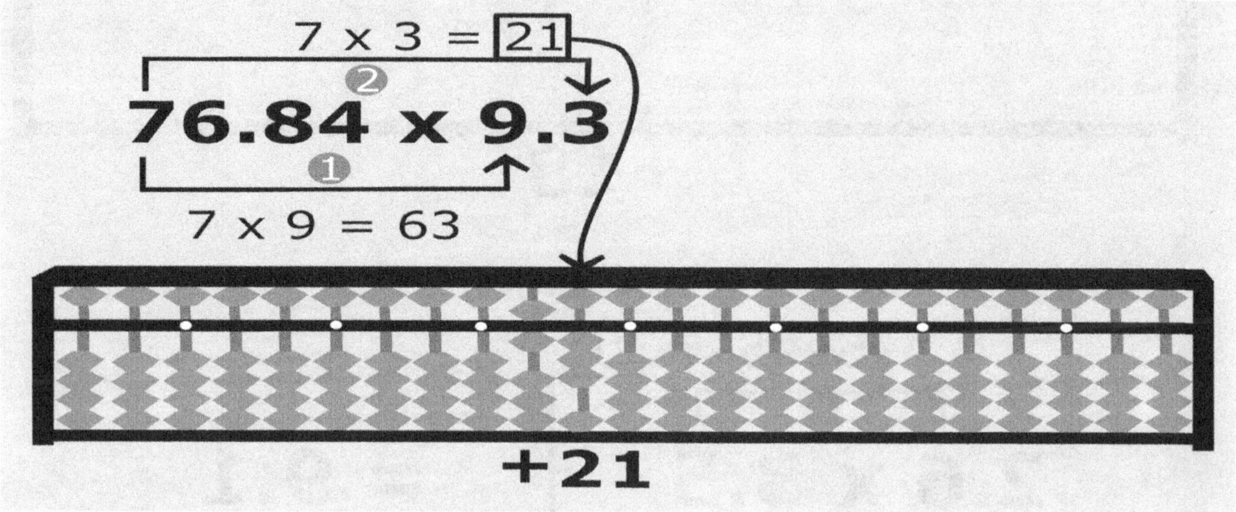
7 x 3 = 21
2
76.84 x 9.3
1
7 x 9 = 63
+21

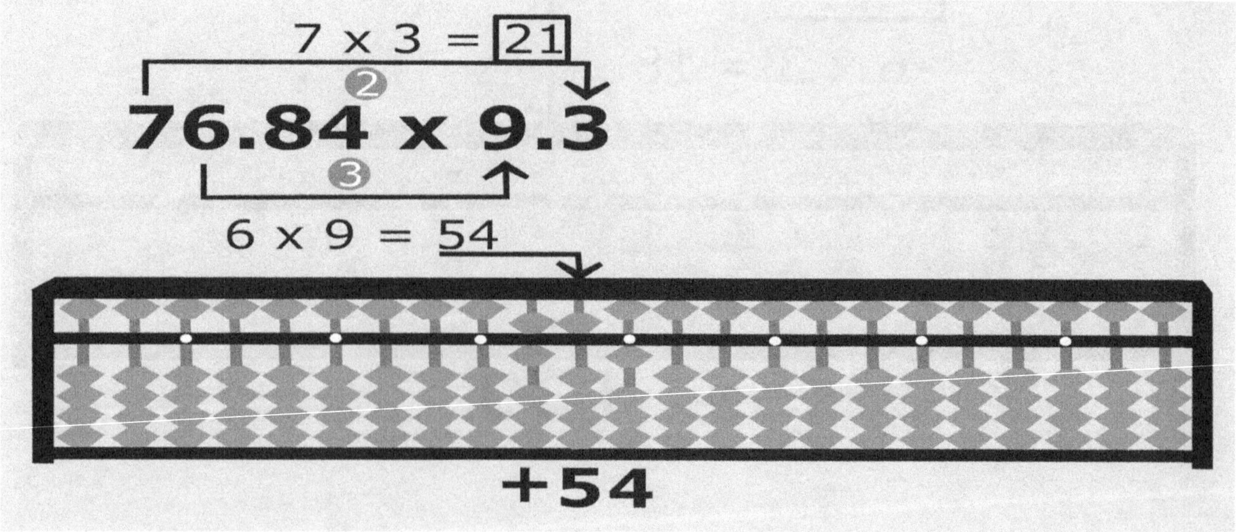
7 x 3 = 21
2
76.84 x 9.3
3
6 x 9 = 54
+54

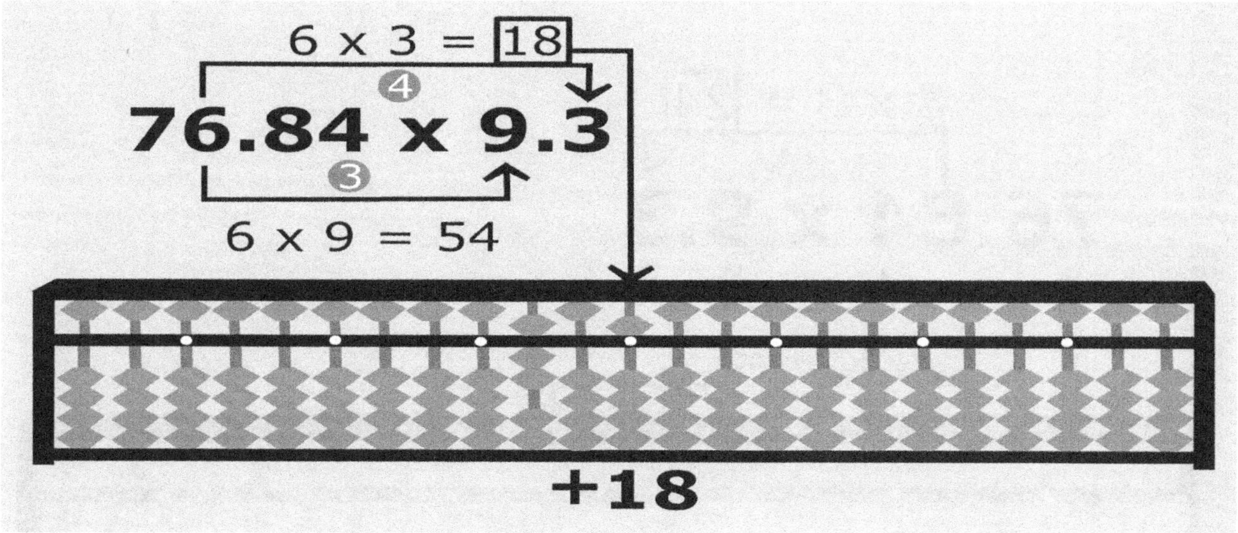
6 x 3 = 18
4
76.84 x 9.3
3
6 x 9 = 54
+18

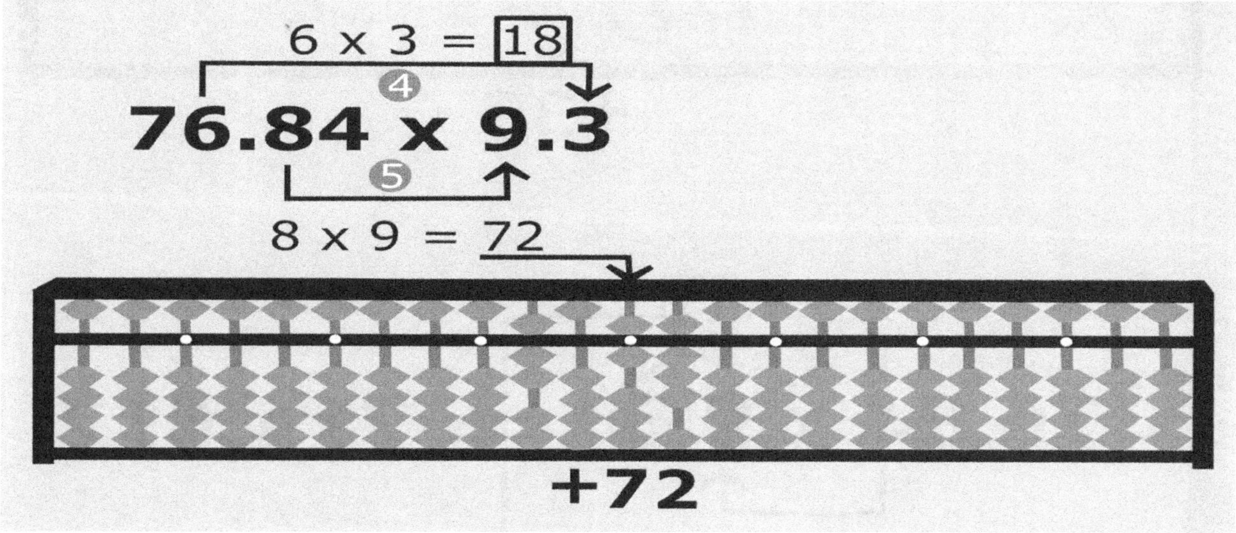
6 x 3 = 18
4
76.84 x 9.3
5
8 x 9 = 72
+72

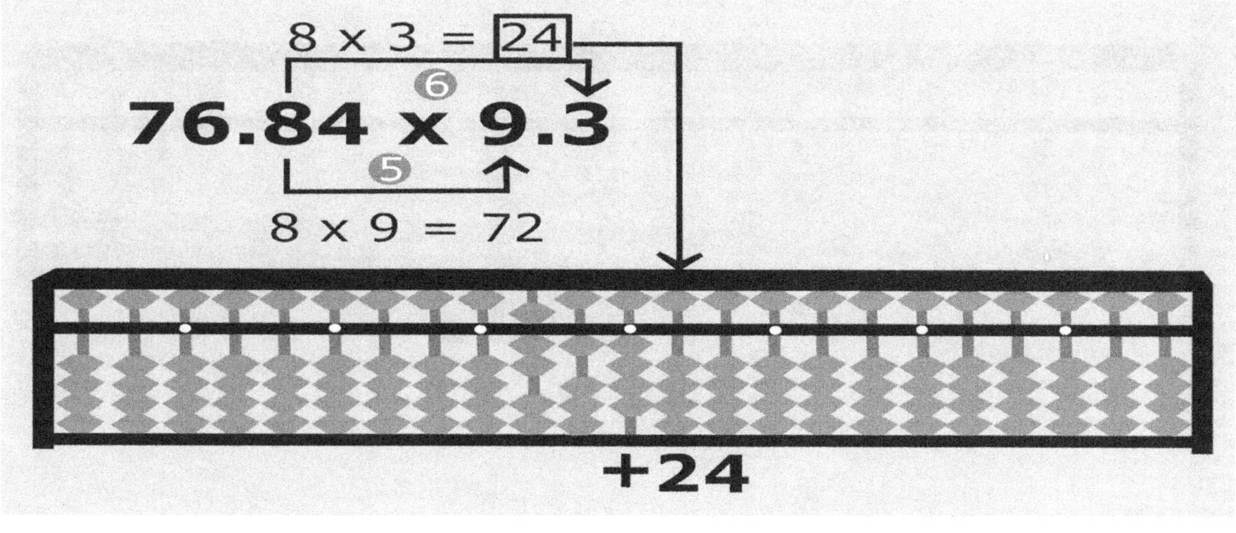
8 x 3 = 24
6
76.84 x 9.3
5
8 x 9 = 72
+24

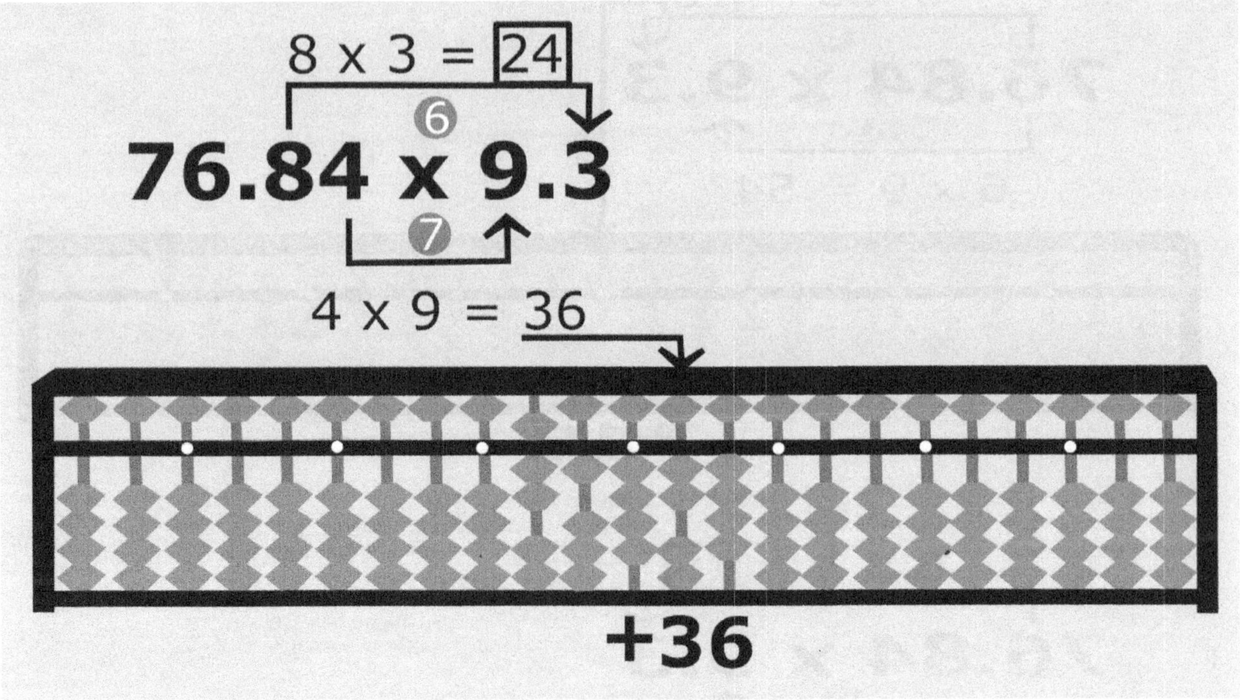
8 x 3 = 24
6
76.84 x 9.3
7
4 x 9 = 36
+36

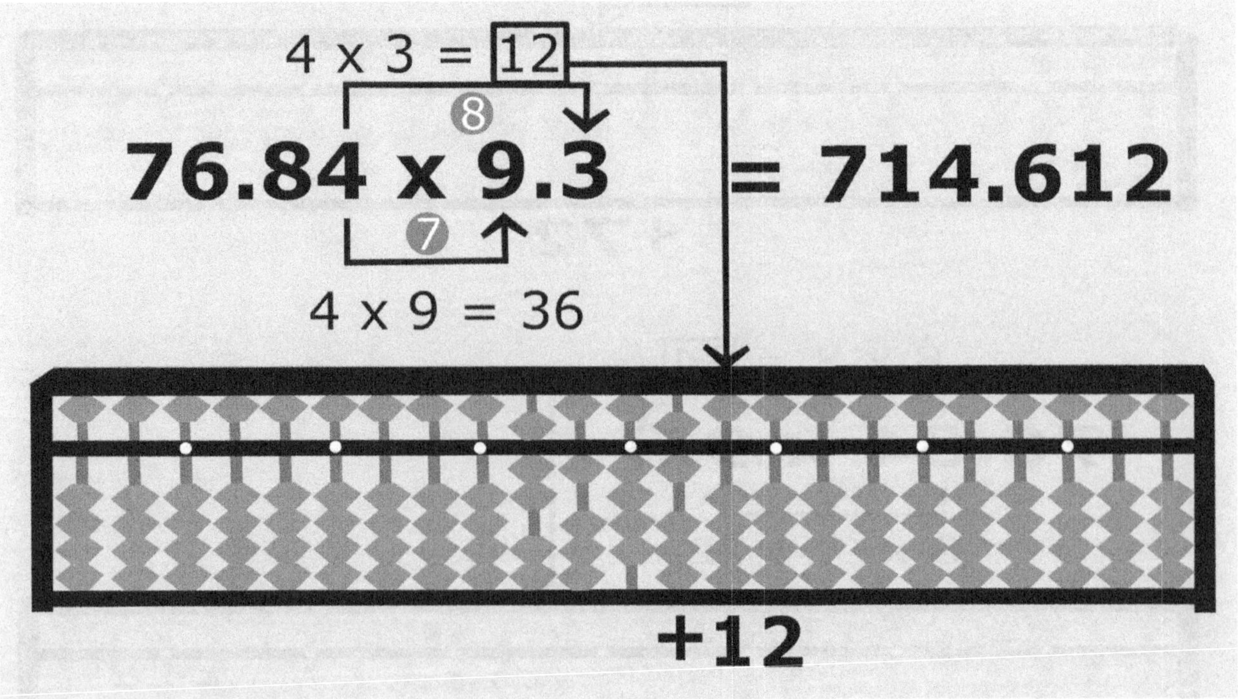
4 x 3 = 12
8
76.84 x 9.3
= 714.612
7
4 x 9 = 36
+12

Practice for decimal multiplication

1) 71 X 2.7 =	15) 84.6 X 3.8 =
2) 95 X 99.6 =	16) 54.7 X 4.8 =
3) 72.3 X 55.4 =	17) 65.98 X 8.3 =
4) 44.27 X 49.6 =	18) 34 X 4.6 =
5) 1.37 X 8.2 =	19) 72.2 X 2.9 =
6) 33.2 X 9.8 =	20) 99.4 X 3.5 =
7) 5.24 X 80.7 =	21) 15.3 X 6.5 =
8) 3.8 X 2.7 =	22) 28 X 28.3 =
9) 22.9 X 2.5 =	23) 55.1 X 89.9 =
10) 10.3 X 6.5 =	24) 48.8 X 60.4 =
11) 8.6 X 4.9 =	25) 77.9 X 9.7 =
12) 14 X 5.9 =	26) 51.8 X 8.7 =
13) 26.9 X 98.3 =	27) 95.2 X 7.3 =
14) 80.2 X 63.5 =	28) 20.1 X 3.8 =

Yay.......

You MADE it!

Let's move on to How to Abacus 4: All About Division

Answer Key

p. 21 (2 digits times 1 digit)

1	2	3	4	5	6	7	8	9	10
276	**588**	**520**	**567**	**252**	**198**	**512**	**222**	**232**	**48**
11	12	13	14	15	16	17	18	19	20
348	**342**	**456**	**316**	**208**	**441**	**465**	**225**	**552**	**188**
21	22	23	24	25	26	27	28		
290	**190**	**360**	**376**	**210**	**792**	**752**	**207**		

p. 25 (3 digits times 1 digit)

1	2	3	4	5	6	7	8	9	10
5,754	**966**	**5,172**	**2,247**	**1,048**	**856**	**3,141**	**5,864**	**2,652**	**1,377**
11	12	13	14	15	16	17	18	19	20
2,040	**828**	**1,695**	**905**	**1,729**	**5,790**	**1,444**	**4,797**	**3,410**	**5,598**
21	22	23	24	25	26	27	28		
570	**3,324**	**2,436**	**925**	**1,677**	**5,352**	**4,110**	**2,296**		

p. 31 (2 digits times 2 digits)

1	2	3	4	5	6	7	8	9	10
1,886	**570**	**624**	**1,343**	**7,200**	**1,095**	**3,132**	**864**	**4,756**	**4,275**
11	12	13	14	15	16	17	18	19	20
2,408	**8,316**	**494**	**1,625**	**3,479**	**5,096**	**3,900**	**2,068**	**1,144**	**6,256**
21	22	23	24	25	26	27	28		
1,104	**3,564**	**3,936**	**2,175**	**288**	**4,736**	**4,697**	**1,537**		

p. 38 (3 digits times 2 digits)

1	2	3	4	5	6	7	8	9	10
5,962	**48,807**	**54,835**	**41,356**	**41,920**	**61,348**	**8,618**	**10,746**	**17,780**	**70,455**
11	12	13	14	15	16	17	18	19	20
15,232	**8,260**	**33,026**	**50,526**	**33,212**	**40,552**	**5,625**	**15,824**	**25,344**	**17,465**
21	22	23	24	25	26	27	28		
23,296	**7,868**	**6,695**	**74,593**	**28,080**	**74,037**	**49,039**	**7,638**		

p. 51 (decimal multiplication)

1	2	3	4	5	6	7	8	9	10
191.7	**9,462**	**4,005.42**	**2,195.792**	**11.234**	**325.36**	**422.868**	**10.26**	**57.25**	**66.95**
11	12	13	14	15	16	17	18	19	20
2.14	**82.6**	**2,644.27**	**5,092.7**	**321.48**	**262.56**	**547.634**	**156.4**	**209.38**	**347.9**
21	22	23	24	25	26	27	28		
99.45	**792.4**	**4,953.49**	**2,947.52**	**755.63**	**450.66**	**694.96**	**76.38**		

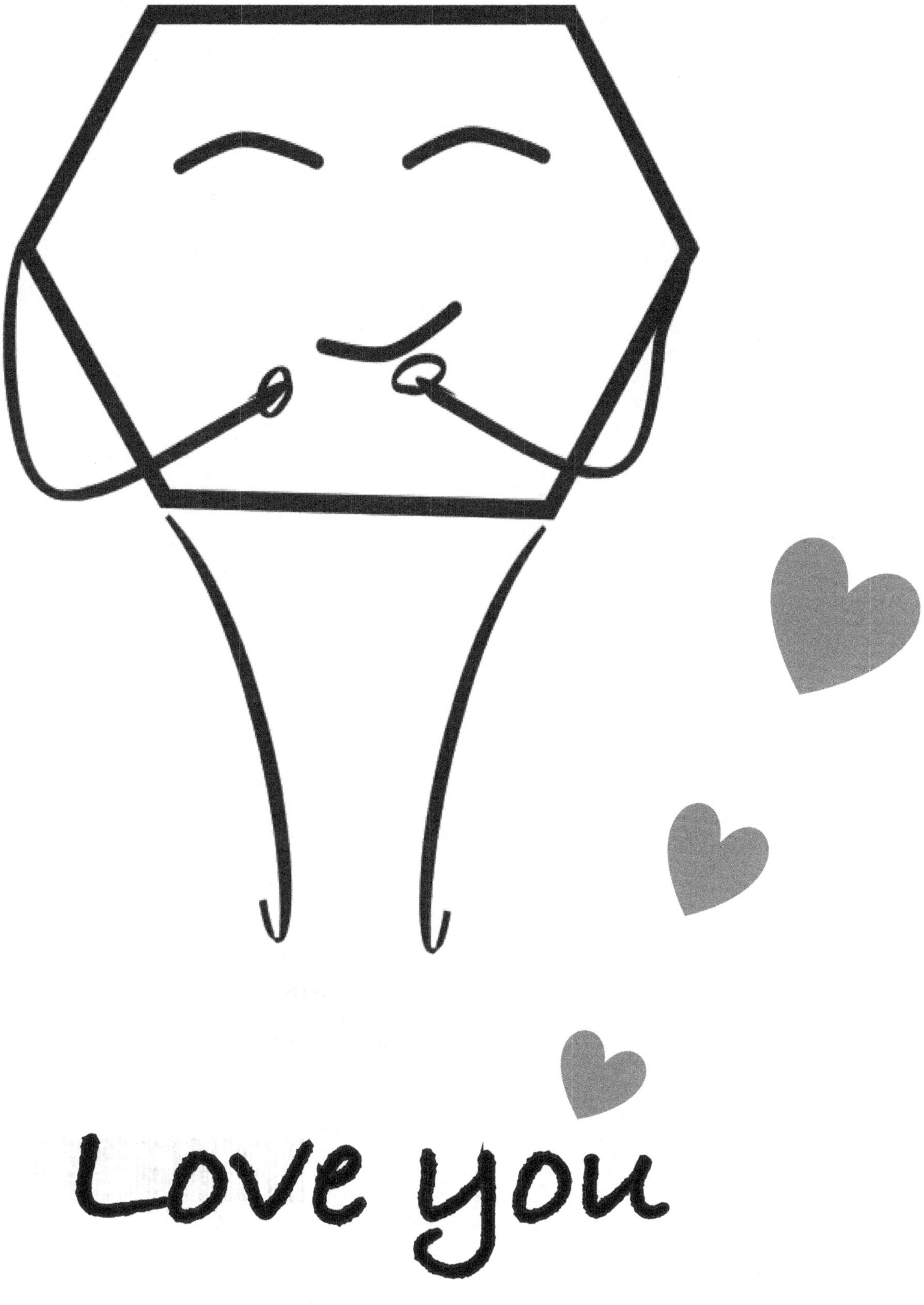
Love you

Made in the USA
Monee, IL
08 November 2026

68085578R00031